Lecture Notes in Social Networks

Lecture Notes in Social Networks (LNSN) comprises volumes covering the theory, foundations and applications of the new emerging multidisciplinary field of social networks analysis and mining. LNSN publishes peer-reviewed works (including monographs, edited works) in the analytical, technical as well as the organizational side of social computing, social networks, network sciences, graph theory, sociology, Semantics Web, Web applications and analytics, information networks, theoretical physics, modeling, security, crisis and risk management, and other related disciplines. The volumes are guest-edited by experts in a specific domain. This series is indexed by DBLP. Springer and the Series Editors welcome book ideas from authors. Potential authors who wish to submit a book proposal should contact Christoph Baumann, Publishing Editor, Springer e-mail: Christoph.Baumann@springer.com

More information about this series at http://www.springer.com/series/8768

Giancarlo Ragozini • Maria Prosperina Vitale
Editors

Challenges in Social Network Research

Methods and Applications

 Springer

Editors
Giancarlo Ragozini
Department of Political Science
University of Naples Federico II
Naples, Italy

Maria Prosperina Vitale
Department of Political and Social Studies
University of Salerno
Fisciano (SA), Italy

ISSN 2190-5428   ISSN 2190-5436   (electronic)
Lecture Notes in Social Networks
ISBN 978-3-030-31465-1   ISBN 978-3-030-31463-7   (eBook)
https://doi.org/10.1007/978-3-030-31463-7

This Springer imprint is published by the registered company Springer Nature Switzerland AG.
The registered company address is: Gewerbestrasse 11, 6330 Cham, Switzerland

Preface

The contributions included in this volume mainly stem from a selection of original papers presented at the Sixth International Workshop on Social Network Analysis, ARS'17, "Challenges in Social network research," held in Naples (Italy) on May 16–17, 2017. The workshop was the sixth of a successful series. It aimed at presenting the most relevant results and the most recent methodological developments in social network research in order to broaden the knowledge of network analysis and discuss "where" and "how" this methodological perspective ought to be applied both in practice and in theory.

ARS ("Analisi delle Reti Sociali", i.e., "Social Network Analysis" in Italian) is a multidisciplinary group of scholars promoting research on social networks by organizing meetings, summer schools, and biennial workshops hosted by Italian universities, as well as by publishing dedicated special issues on the topic since 2007 celebrating the 10-year Edition in 2017.

The book includes both invited and contributed chapters dealing with advanced methods and theoretical development for the analysis of social networks and applications in numerous disciplines. Some authors explore new trends related to network measures, multilevel networks, and clustering on networks, while other contributions deepen the relationship among statistical methods for data mining and social network analysis.

Along with the new methodological developments, the book offers interesting applications to a wide set of fields, ranging from the organizational and economic studies, collaboration, and innovation to the less usual field of poetry. In addition, the case studies are related to local context, showing how the substantive reasoning is fundamental in social network analysis.

The list of authors includes both top scholars in the field of social networks and promising young researchers.

More specifically, the volume is introduced by two invited chapters. In the first chapter, *V. Batagelj* addresses the issues related to two network measures: the overlap weight of an edge and the clustering coefficient of a node proposing a corrected measure of both. The second chapter by *E. Lazega* presents a theoretical contribution on the role of the analysis of multilevel networks in the fields of politics, institutional entrepreneurship, and social change.

The methodological and theoretical parts are opened by a chapter that discusses a novel methodology meant to derive networks and communities from socio-cultural data, based on socio-cultural cognitive mapping and k-NN network modularity maximization, authored by *I. Cruickshank* and *K.M. Carley*. The fourth chapter by *C. Drago* and *R. Ricciuti* presents a new method for the detection of changes in the network structure by a bootstrap test on the degree distribution. The analytical procedure is illustrated by using simulated data and a real-world example on interlocking directorship network. A two-step strategy of analysis is discussed in the chapter by *G. Giordano et al.* for studying comorbidity patterns exploiting association rules extracted by two-mode networks of prescriptions by pathologies and partitioning algorithms to identify the most relevant and connected parts in the one-mode network of pathologies. The sixth chapter by *I. Gollini* investigates the latent structure of bipartite networks via a model-based clustering approach. The model is exploited to identify the latent groups of terrorists and their latent trait scores based on their attendance to certain events.

The chapter by *C. Pinto* proposes a network formation model based on Data Envelopment Analysis in which the usefulness of the agents depends exclusively on direct links and their formation. *M. Schoenfeld* and *J. Pfeffer* emphasize, in the eighth chapter, how network analysis benefits from considering context information and they identify the key challenges that need to be tackled from an algorithmic perspective.

Along with the theoretical and methodological contributions, this volume also offers applications in: conservation agriculture and sustainability, urban community hubs, cultural network and market, policy design and evaluation, Third Sector and inter-organizational networks.

In the ninth chapter, *J. M. Aguirre-López et al.* analyze the adoption patterns of conservation agriculture practices among a sample of maize smallholder farmers in the Mexican state of Chiapas. In the tenth chapter, after a preliminary investigation aimed at collecting a list of active organizations in Naples, *Napolitano et al.* verify and map the correspondence of activities performed by organizations with the priority themes of the cultural production system defined by the European Urban Agenda. In the eleventh chapter, *S. Pedrini* and *C. Felaco* explore the social dynamics and the interactions between young poets and the Italian poetic network with their cultural path by adopting Multiple Correspondence Analysis for two-mode networks. In the twelfth chapter by *M. Russo et al.*, the authors analyze the extent to which innovation intermediaries support the creation of communities of other agents through their engagement in different activities. Multilayer network analysis techniques are adopted to simultaneously represent the several types of interactions promoted by intermediaries. The book is closed by *A. Salvini et al.*, who provide insights on the network governance of the social interacting organizations, presenting an empirical study of two networks of organizations operating in local territories in Southern Italy and focusing on Third Sector and welfare activities.

The editors would like to dedicate this book to professors Maria Rosaria D'Esposito and Anuška Ferligoj, who have been the soul of the ARS group and our personal inspiring mentors in the research activities in this field over our whole academic life and are going to lead us in the future years to come as well.

We would also like to thank the members of the Scientific Program Committee for contributing to make our conference successful.

Special thanks are also due to the anonymous reviewers who helped us in the selection process of the papers.

Naples, Italy   Giancarlo Ragozini
Fisciano, Italy   Maria Prosperina Vitale

Contents

Part II   Applications

Contributors

Juan Manuel Aguirre-López Universidad Autónoma Chapingo, Chapingo, Mexico

Vladimir Batagelj Department of Theoretical Computer Science, Institute of Mathematics, Physics and Mechanics, Jadranska, Ljubljana, Slovenia
University of Primorska, Andrej Marušič Institute, Koper, Slovenia
National Research University Higher School of Economics, Moscow, Russia

Annalisa Caloffi Department of Economics and Management, University of Florence, Florence, Italy

Kathleen M. Carley Center for Computational Analysis of Social and Organizational Systems (CASOS), Institute for Software Research, Carnegie Mellon University, Pittsburgh, PA, USA

Pierpaolo Cavallo Department of Physics E.R. Caianiello, University of Salerno, Fisciano (SA), Italy

Petra Chaloupková Czech University of Life Sciences, Prague, Czech Republic

Iain Cruickshank Center for Computational Analysis of Social and Organizational Systems (CASOS), Institute for Software Research, Carnegie Mellon University, Pittsburgh, PA, USA

Mario De Santis Cooperativa Medi Service, Salerno, Italy

Julio Díaz-José Tecnológico Nacional de México-Campus Zongolica, Veracruz, Mexico
Facultad de Ciencias Biológicas y Agropecuarias, Universidad Veracruzana, Veracruz, Mexico

Carlo Drago Niccolò Cusano University, Rome, Italy

Cristiano Felaco University of Naples Federico II, Naples, Italy

Giuseppe Giordano Department of Political and Social Studies, University of Salerno, Fisciano (SA), Italy

Isabella Gollini University College Dublin, Belfield, Dublin, Ireland

Francisco Guevara-Hernández Universidad Autónoma de Chiapas, Villaflores, Chiapas, México

Emmanuel Lazega Sciences Po, CSO-CNRS, IUF, Paris, France

Pasquale Napolitano Science and Technology for Information and Communication Society, Graphics, Vision, Multimedia IRISS/CNR, Naples, Italy

Sergio Pagano Department of Physics E.R. Caianiello, University of Salerno, Fisciano (SA), Italy

Sabrina Pedrini University of Bologna Alma Mater Studiorum, Bologna, Italy

Juergen Pfeffer Bavarian School of Public Policy, Technical University in Munich, Munich, Germany

Claudio Pinto University of Salerno, Fisciano (SA), Italy

Irene Psaroudakis University of Pisa, Pisa, Italy

Giancarlo Ragozini Department of Political Science, University of Naples Federico II, Naples, Italy

Antonietta Riccardo University of Pisa, Pisa, Italy

Roberto Ricciuti University of Verona, Verona, Italy

Riccardo Righi European Commission, Joint Research Centre (JRC), Seville, Spain

Simone Righi Department of Computer Science, University College London, London, UK
"Lendület" Research Center for Education and Network Studies (RECENS), Hungarian Academy of Sciences, Budapest, Hungary

Federica Rossi Birkbeck, University of London, London, UK

Margherita Russo Department of Economics, University of Modena and Reggio Emilia, Modena, Italy

Andrea Salvini University of Pisa, Pisa, Italy

Mirco Schoenfeld Bavarian School of Public Policy, Technical University in Munich, Munich, Germany

Francesco Vasca University of Sannio, Benevento, Italy

Rita Lisa Vella Department of Business and Management, LUISS Guido Carli, Rome, Italy

Maria Prosperina Vitale Department of Political and Social Studies, University of Salerno, Fisciano (SA), Italy

Pierluigi Vitale Department of Political and Communication Sciences, University of Salerno, Fisciano (SA), Italy

Corrected Overlap Weight
and Clustering Coefficient

Vladimir Batagelj

Abstract We discuss two well-known network measures: the overlap weight of an edge and the clustering coefficient of a node. For both of them it turns out that they are not very useful for data analytic task to identify important elements (nodes or links) of a given network. The reason for this is that they attain their largest values on maximal subgraphs of relatively small size that are more probable to appear in a network than that of larger size. We show how the definitions of these measures can be corrected in such a way that they give the expected results. We illustrate the proposed corrected measures by applying them to the US Airports network using the program Pajek.

Keywords Social network analysis · Importance measure · Triangular weight · Overlap weight · Clustering coefficient

1 Introduction

1.1 Network Element Importance Measures

To identify important/interesting elements (nodes, links) in a network we often try to express our intuition about their importance using an appropriate measure (node index, link weight) following the scheme:

Larger is the measure value of an element, more important/interesting is this element.

Mathematics Subject Classification 2010: 91D30, 91C05, 05C85, 68R10, 05C42

V. Batagelj (✉)
Department of Theoretical Computer Science, Institute of Mathematics, Physics and Mechanics, Jadranska, Ljubljana, Slovenia

University of Primorska, Andrej Marušič Institute, Koper, Slovenia
National Research University Higher School of Economics, Moscow, Russia
e-mail: vladimir.batagelj@fmf.uni-lj.si

Too often, in the analysis of networks, researchers uncritically pick some measure from the literature (degrees, closeness, betweenness, hubs and authorities, clustering coefficient, etc. [1, 2]) and apply it to their network.

In this paper we discuss two well-known network local density measures: the overlap weight of an edge [3] and the clustering coefficient of a node [4, 5].

For both of them it turns out that they are not very useful for data analytic task to identify important elements of a given network. The reason for this is that they attain their largest values on maximal subgraphs of relatively small size—they are more probable to appear in a network than that of larger size. We show how their definitions can be corrected in such a way that they give the expected results. We illustrate the proposed corrected measures by applying them to the US Airports network using the program Pajek. We will limit our attention to undirected simple graphs $\mathbf{G} = (\mathcal{V}, \mathcal{E})$.

Many similar indices and weights were proposed by graph drawing community for disentanglement in the visualization of hairball networks [6–8].

When searching for important subnetworks in a given network we often assume a model that in the evolution of the network the increased activities in a part of the network create new nodes and edges in that part increasing its local density. We expect from a *local density* measure $ld(x, \mathbf{G})$ for an element (node/link) x of network $\mathbf{G}$ the following properties:

ld1.   adding an edge, e, to the local neighborhood, $\mathbf{G}^{(1)}$, does not decrease the local density $ld(x, \mathbf{G}) \leq ld(x, \mathbf{G} \cup e)$.

ld2.   normalization: $0 \leq ld(x, \mathbf{G}) \leq 1$.

ld3. $ld(x, \mathbf{G})$ can attain value 1, $ld(x, \mathbf{G}) = 1$, on the largest subnetwork of certain type in the network.

2   Overlap Weight

2.1   Overlap Weight

A direct measure of the overlap of an edge $e = (u : v) \in \mathcal{E}$ in an undirected simple graph $\mathbf{G} = (\mathcal{V}, \mathcal{E})$ is the number of common neighbors of its end nodes u and v (see Fig. 1). It is equal to $t(e)$—the *number of triangles* (cycles of length 3) to which the edge e belongs. The *edge neighbors subgraph* is labeled $T(\deg(u) - t(e) - 1, t(e), \deg(v) - t(e) - 1)$—the subgraph in Fig. 1 is labeled $T(4, 5, 3)$. There are two problems with this measure:

- it is not normalized (bounded to [0, 1]);
- it does not consider the "potentiality" of nodes u and v to form triangles—there are

$$\min(\deg(u), \deg(v)) - 1 - t(e)$$

nodes in the smaller set of neighbors that are not in the other set of neighbors.

Fig. 1 Neighbors of $e(u : v)$

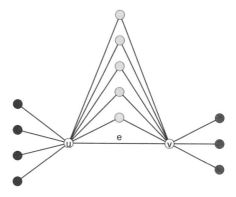

Two simple normalizations are

$$\frac{t(e)}{n-2} \quad \text{or} \quad \frac{t(e)}{\mu},$$

where $n = |\mathcal{V}|$ is the number of nodes, and $\mu = \max_{e\in\mathcal{E}} t(e)$ is the maximum number of triangles on an edge in the graph **G**.

The (topological) *overlap weight* of an edge $e = (u : v) \in \mathcal{E}$ considers also the degrees of edge's end nodes and is defined as

$$o(e) = \frac{t(e)}{(\deg(u) - 1) + (\deg(v) - 1) - t(e)}.$$

In the case $\deg(u) = \deg(v) = 1$ we set $o(e) = 0$. It somehow resolves both problems.

The overlap weight is essentially a Jaccard similarity index [9]

$$J(X, Y) = \frac{|X \cap Y|}{|X \cup Y|}$$

for $X = N(u) \setminus \{v\}$ and $Y = N(v) \setminus \{u\}$, where $N(z)$ is the set of neighbors of a node z. In this case we have $|X \cap Y| = t(e)$ and

$$|X \cup Y| = |X| + |Y| - |X \cap Y| = (\deg(u) - 1) + (\deg(v) - 1) - t(e).$$

Note also that $h(X, Y) = 1 - J(X, Y) = \frac{|X \oplus Y|}{|X \cup Y|}$ is the normalized Hamming distance [9]. The operation $\oplus$ denotes the symmetric difference $X \oplus Y = (X \cup Y) \setminus (X \cap Y)$.

Another normalized overlap measure is the *overlap index* [9]

$$O(e) = O(X, Y) = \frac{|X \cap Y|}{\max(|X|, |Y|)} = \frac{t(e)}{\max(\deg(u), \deg(v)) - 1}.$$

Both measures J and O, applied to networks, have some nice properties. For example: a pair of nodes u and v are structurally equivalent iff $J(X, Y) = O(X, Y) = 1$. Therefore the overlap weight measures the *substitutiability* of one edge's end node by the other.

Introducing two auxiliary quantities

$$m(e) = \min(\deg(u), \deg(v)) - 1 \quad \text{and} \quad M(e) = \max(\deg(u), \deg(v)) - 1$$

we can rewrite the definition of the overlap weight

$$o(e) = \frac{t(e)}{m(e) + M(e) - t(e)}, \quad M(e) > 0$$

and if $M(e) = 0$, then $o(e) = 0$.

For every edge $e \in \mathcal{E}$ it holds $0 \leq t(e) \leq m(e) \leq M(e)$. Therefore

$$m(e) + M(e) - t(e) \geq t(e) + t(e) - t(e) = t(e)$$

showing that $0 \leq o(e) \leq 1$.

The value $o(e) = 1$ is attained exactly in the case when $M(e) = t(e)$, and the value $o(e) = 0$ exactly when $t(e) = 0$.

In simple directed graphs without loops different types of triangles exist over an arc $a(u, v)$. We can define overlap weights for each type. For example: the *transitive overlap weight*

$$o_t(a) = \frac{t_t(a)}{(\operatorname{outdeg}(u) - 1) + (\operatorname{indeg}(v) - 1) - t_t(a)}$$

and the *cyclic overlap weight*

$$o_c(a) = \frac{t_c(a)}{\operatorname{indeg}(u) + \operatorname{outdeg}(v) - t_c(a)},$$

where $t_t(a)$ and $t_c(a)$ are the number of transitive/cyclic triangles containing the arc a. In this paper we will limit our discussion to overlap weights in undirected graphs.

2.2 US Airports Links with the Largest Overlap Weight

Let us apply the overlap weight to the network of US Airports 1997 [10]. It consists of 332 airports and 2126 edges among them. There is an edge linking a pair of airports iff in the year 1997 there was a flight company providing flights between those two airports.

The size of a circle representing an airport in Fig. 2 is proportional to its degree—the number of airports linked to it. The airports with the largest degree are:

Airport	deg
Chicago O'hare Intl	139
Dallas/Fort Worth Intl	118
The William B Hartsfield Atlanta	101
Lambert-St Louis Intl	94
Pittsburgh Intl	94

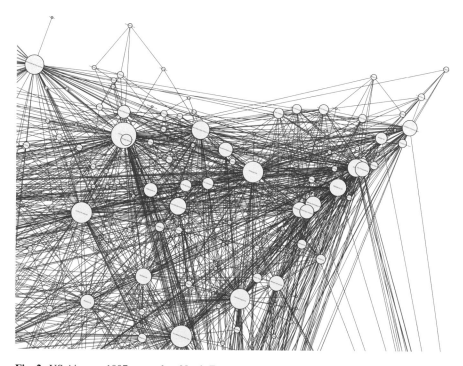

Fig. 2 US Airports 1997 network, a North-East cut-out

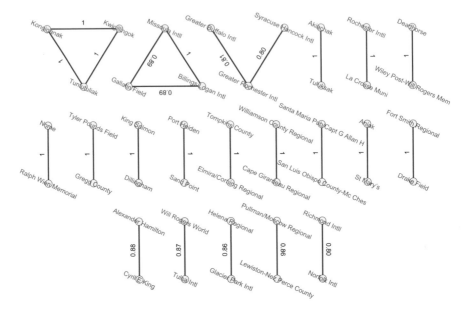

Fig. 3 Edges with the largest overlap—cut at 0.8

Fig. 4 Zoom in a tetrahedron
(Kwigillingok, Kongiganak,
Tuntutuliak, Bethel)

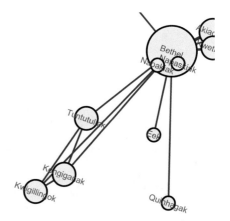

For the overlap weight the edge cut at level 0.8 (a subnetwork of all edges with overlap weight at least 0.8) is presented in Fig. 3. It consists of two triangles, a path of length 2, and 17 separate edges.

A tetrahedron (Kwigillingok, Kongiganak, Tuntutuliak, Bethel), see Fig. 4, gives the first triangle in Fig. 3—attached with the node Bethel to the rest of the network.

From this example we see that in real-life networks edges with the largest overlap weight tend to be edges with relatively small degrees in their end nodes ($o(e) = 1$ implies $\deg(u) = \deg(v) = t(e) + 1$)—the overlap weight does not satisfy the condition **ld3**. Because of this the overlap weight is not very useful for data analytic tasks in searching for important elements of a given network. We would like to emphasize here that there are many applications in which the overlap weight proves to be useful and appropriate; we question only its appropriateness for determining the most overlapped edges. We will try to improve the overlap weight definition to better suit the data analytic goals.

2.3 Corrected Overlap Weight

We define a *corrected overlap weight* as

$$o'(e) = \frac{t(e)}{\mu + M(e) - t(e)}.$$

By the definition of μ for every $e \in \mathcal{E}$ it holds $t(e) \le \mu$. Since $M(e) - t(e) \ge 0$ also $\mu + M(e) - t(e) \ge \mu$ and therefore **ld2**, $0 \le o'(e) \le 1$. $o'(e) = 0$ exactly when $t(e) = 0$, and $o'(e) = 1$ exactly when $\mu = M(e) = t(e)$. For **ld3**, the corresponding maximal edge neighbors subgraph contains $T(0, \mu, 0)$. The end nodes of the edge e are structurally equivalent.

To show that **ld1** also holds let $\mathbf{G}^{(1)}(e)$ denote the edge neighbors subgraph of the edge e. Let f be an edge added to $\mathbf{G}^{(1)}(e)$. We can assume that $\deg(u) \ge \deg(v)$, $e = (u : v)$. Therefore $M(e) = \deg(u) - 1$. We have to consider some cases:

a. $f \in \mathcal{E}(\mathbf{G}^{(1)}(e))$: then $\mathbf{G} \cup f = \mathbf{G}$ and $o'(e, \mathbf{G} \cup f) = o'(e, \mathbf{G})$.
b. $f \notin \mathcal{E}(\mathbf{G}^{(1)}(e))$:

> **b1.** $f = (u : t)$: then $t \in N(v) \setminus T(e) \setminus e$. It creates new triangle (u, v, t). We have $t'(e) = t(e) + 1$ and $M'(e) = M(e) + 1$. We get

$$o'(e, \mathbf{G} \cup f) = \frac{t'(e)}{\mu + M'(e) - t'(e)} = \frac{t(e) + 1}{\mu + M(e) - t(e)} > o'(e, \mathbf{G})$$

> **b2.** $f = (v : t)$: then $t \in N(u) \setminus T(e) \setminus e$. It creates new triangle (u, v, t). We have $t'(e) = t(e) + 1$ and $M'(e) = M(e)$. We get

$$o'(e, \mathbf{G} \cup f) = \frac{t'(e)}{\mu + M'(e) - t'(e)} = \frac{t(e) + 1}{\mu + M(e) - t(e) - 1}$$
$$> \frac{t(e) + 1}{\mu + M(e) - t(e)} > o'(e, \mathbf{G})$$

> **b3.** $f = (t : w)$ and $t, w \in N(u) \cup N(v) \setminus \{u, v\}$: No new triangle on e is created. We have $t'(e) = t(e)$ and $M'(e) = M(e)$. Therefore $o'(e, \mathbf{G} \cup f) = o'(e, \mathbf{G})$.

The corrected overlap weight o' is a kind of local density measure, but it is primarily a substitutiability measure. To get a better local density measure we have to consider besides triangles also quadrilaterals (4-cycles).

2.4 US Airports 1997 Links with the Largest Corrected Overlap Weight

For the US Airports 1997 network we get $\mu = 80$. For the corrected overlap weight the edge cut at level 0.5 is presented in Fig. 5. Six links with the largest triangular weights are given in Table 1.

In Fig. 6 all the neighbors of end nodes WB Hartsfield Atlanta and Charlotte/Douglas Intl of the link with the largest corrected overlap weight value are presented. They have 76 common (triangular) neighbors. The node WB Hartsfield Atlanta has 11 and the node Charlotte/Douglas Intl has 25 additional neighbors. Note (see Table 1) that there are some links with higher triangular weight, but also with a much higher number of additional neighbors—therefore with smaller corrected overlap weights.

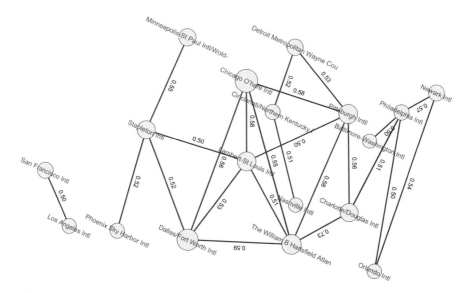

Fig. 5 US Airports 1997 links with the largest corrected overlap weight, cut at 0.5

Table 1 Largest triangular weights in US Airports 1997 network

u	v	$t(e)$	$d(u)$	$d(v)$	$o'(e)$
Chicago O'hare Intl	Pittsburgh Intll	80	139	94	0.57971
Chicago O'hare Intl	Lambert-St Louis Intl	80	139	94	0.57971
Chicago O'hare Intl	Dallas/Fort Worth Intl	78	118	139	0.55714
Chicago O'hare Intl	The W B Hartsfield Atlanta	77	101	139	0.54610
The W B Hartsfield Atlanta	Charlotte/Douglas Intl	76	101	87	0.73077
The W B Hartsfield Atlanta	Dallas/Fort Worth Intl	73	101	118	0.58871

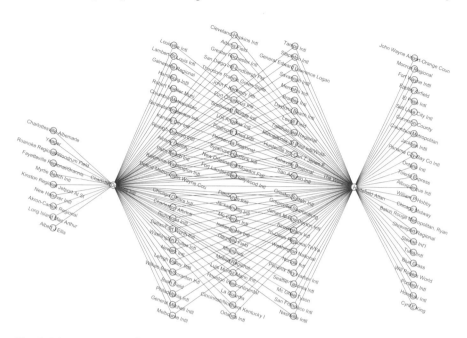

Fig. 6 US Airports links o'(WB Hartsfield Atlanta, Charlotte/Douglas Intl) = 0.7308

2.5 Comparisons

In Fig. 7 the set $\{(o(e), o'(e)) : e \in \mathcal{E}\}$ is displayed for the US Airports 1997 network. For most edges it holds $o'(e) \leq o(e)$. It is easy to see that $o(e) < o'(e) \Leftrightarrow \mu < m(e)$. Edges with the overlap value $o(e) > 0.8$ have the corrected overlap weight $o'(e) < 0.2$.

In Fig. 8 the sets $\{(m(e), o(e)) : e \in \mathcal{E}\}$ and $\{(m(e), o'(e)) : e \in \mathcal{E}\}$ are displayed for the US Airports 1997 network. With increasing of $m(e)$ the corresponding overlap weight $o(e)$ is decreasing, and the corresponding corrected overlap weight $o'(e)$ is also increasing.

We can observe similar tendencies if we compare both weights with respect to the number of triangles $t(e)$ (see Fig. 9).

3 Clustering Coefficient

3.1 Clustering Coefficient

For a node $u \in \mathcal{V}$ in an undirected simple graph $\mathbf{G} = (\mathcal{V}, \mathcal{E})$ its (local) *clustering coefficient* [9] is measuring a local density in the node u and is defined as a

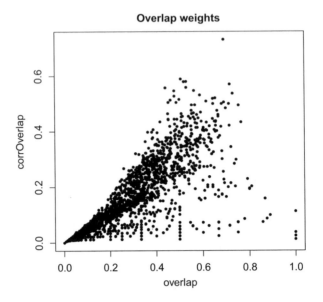

Fig. 7 Comparison (overlap, corrected overlap)

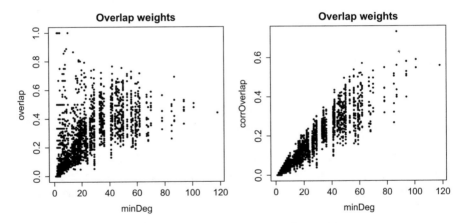

Fig. 8 Comparison—minDeg(e)

proportion of the number of existing edges between u's neighbors to the number of all possible edges between u's neighbors

$$cc(u) = \frac{|\mathcal{E}(N(u))|}{|\mathcal{E}(K_{\deg(u)})|} = \frac{2 \cdot E(u)}{\deg(u) \cdot (\deg(u) - 1)}, \quad \deg(u) > 1,$$

where $E(u) = |\mathcal{E}(N(u))|$. If $\deg(u) \leq 1$, then $cc(u) = 0$.

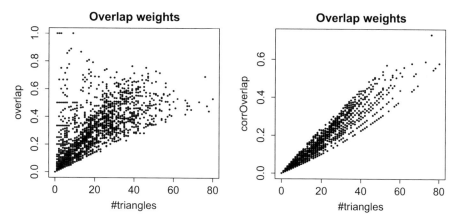

Fig. 9 Comparison—# of triangles

It is easy to see that

$$E(u) = \frac{1}{2} \sum_{e \in S(u)} t(e),$$

where $S(u) = \{e(u : v) : e \in \mathcal{E}\}$ is the star in node u.

It holds $0 \le cc(u) \le 1$; $cc(u) = 1$ exactly when $\mathcal{E}(N(u))$ is isomorphic to $K_{\deg(u)}$—a complete graph on $\deg(u)$ nodes. Therefore it seems that the clustering coefficient could be used to identify nodes with the densest neighborhoods.

The notion of clustering coefficient can be extended also to simple directed graphs (with loops).

3.2 US Airports with the Largest Clustering Coefficient

Let us apply also the clustering coefficient to the US Airports 1997 network.

In Table 2 airports with the clustering coefficient equal to 1 and the degree at least 4 are listed. There are 28 additional such airports with a degree 3 and 38 with a degree 2.

Again we see that the clustering coefficient attains its largest value in nodes with a relatively small degree. The probability that wc get a complete subgraph on $N(u)$ is decreasing very fast with increasing of $\deg(u)$. The clustering coefficient does not satisfy the condition **ld3**.

Table 2 US Airports 1997 with clustering coefficient $= 1$

n	deg	Airport	n	deg	Airport
1	7	Lehigh Valley Intll	8	4	Gunnison County
2	5	Evansville Regional	9	4	Aspen-Pitkin Co/Sardy Field
3	5	Stewart Int'l	10	4	Hector Intll
4	5	Rio Grande Valley Intl	11	4	Burlington Regional
5	5	Tallahassee Regional	12	4	Rafael Hernandez
6	4	Myrtle Beach Intl	13	4	Wilkes-Barre/Scranton Intl
7	4	Bishop Intll	14	4	Toledo Express

3.3   Corrected Clustering Coefficient

To get a corrected version of the clustering coefficient we proposed in Pajek [11] to replace $\deg(u)$ in the denominator with $\Delta = \max_{v \in \mathcal{V}} \deg(v)$. In this paper we propose another solution—we replace $\deg(u) - 1$ with μ:

$$cc'(u) = \frac{2 \cdot E(u)}{\mu \cdot \deg(u)}, \quad \deg(u) > 0.$$

If $\deg(u) = 0$, then $cc'(u) = 0$. Note that, if $\Delta > 0$ then $\mu < \Delta$.

To verify the property **ld1** we add to $\mathbf{G}(u)$ a new edge f with its end nodes in $\mathbf{G}(u)$. Then $E'(u) = E(u) + 1$ and $\deg'(u) = \deg(u)$. Therefore

$$cc'(u, \mathbf{G} \cup f) = \frac{2 \cdot E'(u)}{\mu \cdot \deg'(u)} = \frac{2 \cdot (E(u) + 1)}{\mu \cdot \deg(u)} > cc'(u, \mathbf{G}).$$

To show the property **ld2**, $0 \le cc'(u) \le 1$, we have to consider two cases:

a. $\deg(u) \ge \mu$: then for $v \in N(u)$ we have $\deg_{N(u)}(v) \le \mu$ and therefore

$$2 \cdot E(u) = \sum_{v \in N(u)} \deg_{N(u)}(v) \le \sum_{v \in N(u)} \mu = \mu \cdot \deg(u)$$

b. $\deg(u) < \mu$: then $\deg(u) - 1 \le \mu$ and therefore

$$2 \cdot E(u) \le \deg(u) \cdot (\deg(u) - 1) \le \mu \cdot \deg(u)$$

For the property **ld3**, the value $cc'(u) = 1$ is attained in the case **a** on a μ-core, and in the case **b** on $K_{\mu+1}$.

3.4 US Airports Nodes with the Largest Corrected Clustering Coefficient

In Table 3 US Airports with the largest corrected clustering coefficient are listed. The largest value 0.3739 is attained for Cleveland-Hopkins Intl airport. In Fig. 10 the adjacency matrix of a subnetwork on its 45 neighbors is presented. The subnetwork is relatively complete. A small value of corrected clustering coefficient is due to relatively small deg $= 45$ with respect to $\mu = 80$.

3.5 Comparisons

In Fig. 11 the set $\{(cc(e), cc'(e)) : e \in \mathcal{E}\}$ is displayed for the US Airports 1997 network. The correlation between both coefficients is very small. An important observation is that edges with the largest value of the clustering coefficient have relatively small values of the corrected clustering coefficient. We also see that the number of edges in a node's neighborhood is almost functionally dependent on its degree.

From Fig. 12 we see that the clustering coefficient is decreasing with the increasing degree. Nodes with a large degree have small values of the clustering coefficient. The values of corrected clustering coefficient are large for nodes of large degree.

Table 3 US Airports 1997 with the largest corrected clustering coefficient

Rank	Value	deg	Id
1	0.3739	45	Cleveland-Hopkins Intl
2	0.3700	50	General Edward Lawrence Logan
3	0.3688	56	Orlando Intl
4	0.3595	42	Tampa Intl
5	0.3488	61	Cincinnati/Northern Kentucky Intl
6	0.3457	70	Detroit Metropolitan Wayne County
7	0.3455	67	Newark Intl
8	0.3429	53	Baltimore-Washington Intl
9	0.3415	47	Miami Intl
10	0.3405	42	Washington National
11	0.3379	56	Nashville Intll
12	0.3359	46	John F Kennedy Intl
13	0.3347	62	Philadelphia Intl
14	0.3335	41	Indianapolis Intl
15	0.3335	50	La Guardia

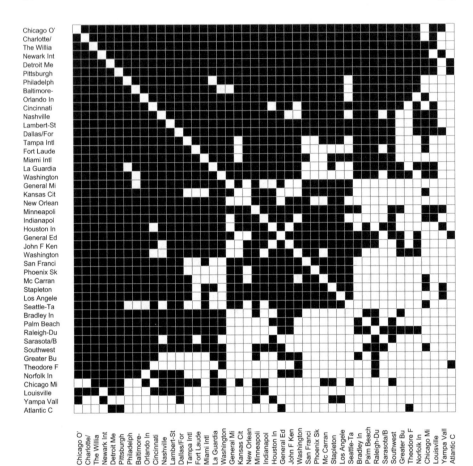

Fig. 10 Links among Cleveland-Hopkins Intl neighbors

4 Conclusions

In this paper we showed that two network measures, the overlap weight and clustering coefficient, are not suitable for the data analytic task of determining important elements in a given network. We proposed corrected versions of these two measures that give expected results.

Because $\mu \leq \Delta$ we can replace in the corrected measures μ with Δ. Its advantage is that it can be easier computed, but the corresponding corrected index is less "sensitive."

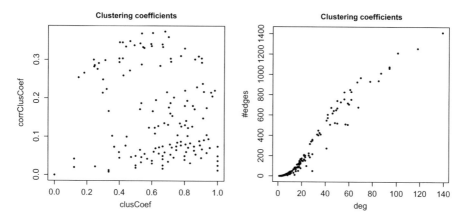

Fig. 11 Comparison—ordinary and corrected clustering coefficients; degrees and number of edges

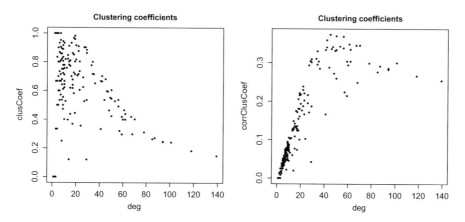

Fig. 12 Comparison—ordinary and corrected clustering coefficients: degrees

An interesting task for future research is a comparison of the proposed measures with measures from graph drawing [6–8].

Acknowledgements The computations were done combining Pajek [11] with short programs in Python and R [12].

This work is supported in part by the Slovenian Research Agency (research program P1-0294 and research projects J1-9187 and J7-8279) and by Russian Academic Excellence Project "5-100."

This paper is a detailed and extended version of the talk presented at the CMStatistics (ERCIM) 2015 Conference. The author's attendance on the conference was partially supported by the COST Action IC1408—CRoNoS.

References

1. Wasserman, S., Faust, K.: Social Network Analysis Methods and Applications. Structural Analysis in the Social Sciences. Cambridge University Press, Cambridge (1995)
2. Todeschini, R., Consonni, V.: Molecular Descriptors for Chemoinformatics, 2nd edn. Wiley-VCH, Weinheim (2009)
3. Onnela, J.P., Saramaki, J., Hyvonen, J., Szabo, G., Lazer, D., Kaski, K., Kertesz, J., Barabasi, A.L.: Structure and tie strengths in mobile communication networks. Proc. Natl. Acad. Sci. **104**(18), 7332 (2007)
4. Holland, P.W., Leinhardt, S.: Transitivity in structural models of small groups. Comp. Group Stud. **2**, 107–124 (1971)
5. Watts, D.J., Strogatz, S.: Collective dynamics of 'small-world' networks. Nature **393**(6684), 440–442 (1998)
6. Melançon, G., Sallaberry, A.: Edge metrics for visual graph analytics: a comparative study. In: 12th International Conference Information Visualisation, pp. 610–615 (2008)
7. Nocaj, A., Ortmann, M., Brandes, U.: Untangling the hairballs of multi-centered, small-world online social media networks. J. Graph Algorithms Appl. **19**(2), 595–618 (2015)
8. Nocaj, A., Ortmann, M., Brandes, U.: Adaptive disentanglement based on local clustering in small-world network visualization. IEEE Trans. Vis. Comput. Graph. **22**(6), 1662–1671 (2016)
9. Wikipedia: (2018). Clustering coefficient: https://en.wikipedia.org/wiki/Clustering_coefficient, Overlap coefficient: https://en.wikipedia.org/wiki/Overlap_coefficient, Hamming distance: https://en.wikipedia.org/wiki/Hamming_distance, Jaccard index: https://en.wikipedia.org/wiki/Jaccard_index
10. Batagelj, V., Mrvar, A.: Pajek data sets: US Airports network (2006). http://vlado.fmf.uni-lj.si/pub/networks/data/mix/USAir97.net
11. De Nooy, W., Mrvar, A., Batagelj, V.: Exploratory Social Network Analysis with Pajek; Revised and Expanded Edition for Updated Software. Structural Analysis in the Social Sciences. Cambridge University Press, Cambridge (2018)
12. Batagelj, V.: Corrected (2016). https://github.com/bavla/corrected

Bottom-Up Collegiality, Top-Down Collegiality, or Inside-Out Collegiality? Analyses of Multilevel Networks, Institutional Entrepreneurship and Laboratories for Social Change

Emmanuel Lazega

Abstract This paper argues that the analysis of multilevel networks (AMN) is useful to understand politics, institutional entrepreneurship, and social change. AMN helps identify multilevel relational infrastructures (in particular multilevel social status) on which institutional entrepreneurship depends, especially in collegial oligarchies as laboratories for social change. In heavily bureaucratized societies, these laboratories take various forms such as bottom-up collegiality, top-down collegiality, and inside-out collegiality. We argue that, in an era of vital transitions, one of the main challenges for social network analyses is to use AMN to observe these collegial oligarchies and to model and understand social (in)capacities to build alternative multilevel relational infrastructures promoting social change. This challenge leads to another: that of understanding the conditions under which a form of collegiality is selected by contextualizing institutional entrepreneurship and its multilevel relational infrastructures. The paper theorizes organized mobility and relational turnover as important dimensions of this contextualization of institution-alization processes.

Keywords Analysis of multilevel networks · Institutional entrepreneurship · Bottom-up collegiality · Top-down collegiality · Inside-out collegiality · Multilevel relational infrastructures · Organized mobility · Multispin

1 Introduction

Historical transitions require new institutions. In this paper we first suggest that analyses of multilevel networks (AMN) provide a new understanding of institutional entrepreneurship. AMN offers models and methods for research designs based

E. Lazega (✉)
Sciences Po, CSO-CNRS, IUF, Paris, France
e-mail: emmanuel.lazega@sciencespo.fr

© Springer Nature Switzerland AG 2020
G. Ragozini, M. P. Vitale (eds.), *Challenges in Social Network Research*,
Lecture Notes in Social Networks, https://doi.org/10.1007/978-3-030-31463-7_2

on linked inter-individual and inter-organizational networks in which each of the superposed networks represents a level of collective agency. Individual members of one network belong to organizations of the other network through affiliation ties. This structural "linked design" [1] extends the sociological concept of duality [2] in which individuals and groups co-constitute each other. Generalizations of this formalism [3] craft a "formal theory of interpenetration" of levels. Articulation of distinct levels of action can be partly accounted for, beyond bipartite structures, with statistical analysis of such datasets [4, 5]. Resources exchanged at each level are of different types. Figure 1 represents a static multilevel network based on this design.

We then show how AMN helps understand institutionalization processes in organized settings by identifying levels of collective agency as either bureaucratic or collegial, and key players or institutional entrepreneurs as active at two (or more) levels of agency simultaneously. They build and maintain multilevel relational infrastructures (MLRIs), in this case multilevel forms of social status. Such political processes and their negotiations are never routine, and therefore necessarily collegial. However, in a bureaucratized organizational society [6], collegiality is always combined with bureaucracy. We identify three multilevel combinations of bureaucracy and collegiality: bottom-up collegiality, top-down collegiality, and inside-out collegiality. Each characterizes a different kind of institutional entrepreneurship in a multilevel context. An example, that of the emergence of a new European institution (the Unified Patent Court), is used to illustrate top-down collegiality in institutional entrepreneurship. In this setting, a collegial oligarchy of judges with multilevel status, i.e., particularly active simultaneously at two levels of agency, i.e., a discrete cluster of "vertical linchpins" who are big fish in big ponds at the national and transnational levels, negotiates and imposes its conception of a new intellectual property regime for European economies.

One of the scientific issues currently challenging social scientists studying social processes in dynamic multilevel networks is that institutional entrepreneurship has determinants and must be contextualized. Building and maintaining MLRIs is not a collective adventure that takes place in a vacuum. We argue that this contextualization must be approached with (and AMN models enriched with data on) at least two determinants of social processes in general: organized mobility of

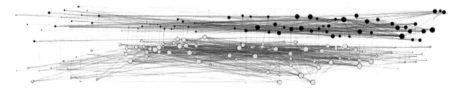

Fig. 1 Big/little fish in big/small ponds: A multilevel network based on linked design. In these superposed networks, white nodes represent individuals, black nodes represent organizations, and ties between white and black nodes are affiliation ties of individuals to organizations. The size of the nodes represents their centrality scores in the network of their level of collective agency

institutional entrepreneurs and relational turnover in their networks (OMRT) at each level of collective agency.

A theory of the effect of OMRT on institutionalization requires intuitions on organized mobility of institutional entrepreneurs and relational turnover in their multilevel networks and inspiration for hypotheses are drawn from the metaphor of the "multispin." In conclusion we argue that, by illuminating (understanding and explaining) the effect of OMRT and its dynamic invariants on the process of institution building, AMN should not only help model social (in)capacities to build new laboratories for social change but also help us understand how to keep this process accountable and democratic.

2   Politics, Analysis of Multilevel Networks, and Multilevel Relational Infrastructures

Interdependencies between actors are too important in social life to be left unorganized, and actors and institutions struggle to organize them. Institutions are among the most venerable objects of study in the social sciences [7]. To simplify, institutions can commonly be defined as rules, norms, and beliefs that describe reality for actors, explaining what is and is not, what can be acted upon and what cannot, and how [8]. Contemporary thinking about the emergence of institutions is dominated in sociology by a variety of neo-institutional perspectives focusing on how norms promoted by institutional entrepreneurs elaborate taken for granted cultural categories, classifications, rules, and procedures that include beliefs and codes stabilizing action into routines [9]. Such a perspective has been shown to lack structure and agency [10–12].

Neo-structural sociology revisits this process by opening it to individual and collective agency, including work of organizing interdependencies. Social network analysis is then used, together with other methods, for tracking and understanding actors' efforts to manage their interdependencies in contexts of cooperation and/or competition where interests diverge, conflicts flare up, constraining but often fragile and polynormative institutions are inherited from the past. As such it avoids reification of the notion of structure and helps in further developing a sociological theory of collective action and of the management of cooperation dilemmas [13–15]. Intentional, reflexive, and strategic behavior endogenizing the structure, not blind reproduction of underlying structure, are parts of the behavioral assumptions of this approach, including the use of organizations as "'tools with a life of their own" and "dynamic conditioning fields" [16], i.e., as political communities in which new institutions are constructed.

For social scientists, finding the links between structure, multilevel position, and collective agency in an organizational society is therefore still a complex task if it has to be carried out in a meaningful way, i.e., in a way that makes normative controversies, conflicts, and politics more intelligible. To do this, it is important to take into account the vertical complexities of the social world. This means differentiating between levels of collective agency and articulating their dynamics in measurements and models. Selznick's "dynamic conditioning field" can thus be considered as a contextual effect and seen as a precursor to contemporary lines of research on multilevel stochastic actor-oriented models [17], multilevel exponential random graph models [5] and multilevel blockmodeling [18–20].

For sociologists accounting for these vertical complexities, rule-making is a complex multilevel political processes in which it is not always easy to identify who is responsible for the promotion of which rule, for example for successes or failures of a transnational regulatory regime. Observations of regulatory activities show that individuals with specific structural characteristics punch above their weight in terms of regulatory activity by precisely being active at two levels of management of interdependencies (advice and contract, for example) at the same time. Presence, participation, and decision-making activity at two or more levels simultaneously allow for cross-level influence at each level separately and, via such "vertical linchpins" [11], jointly. Whether or not such actors are accountable to others in similar ways at different levels, whether or not the rules that they promote are recognized as public goods, are important questions that theory and methodology should help address. Often rules are made discretely, and it takes very sharp stakeholders, experts, non-governmental organizations and journalists to evaluate them, with much regulatory inertia built into the system, much disagreement about whether or not a rule "works" in terms of protecting particular interests—especially the interests of the weakest parties.

In sum the construction and/or maintenance of multilevel relational infrastructure become a step towards coordination within and between levels. Identifying some of the social realities for which multilevel networks are indicators leads to the notions of overlap and complementarity between levels. But it also shows that these levels co-constitute each other via the social construction of MLRIs. These MLRIs are vertical and horizontal differentiations between members (for example forms of multilevel social status of vertical linchpins and multilevel social niches) that are used to influence, from one level, events and processes at other levels. AMN is helpful in showing whether and how MLRI-based complex dynamics and coordination (between individuals, between organizations, and cross-level between individuals and organizations driving each other's evolution) are the laboratories of institutionalization processes and social change in the organizational society.

Dynamics of such multilevel systems of collective agency assume that the evolution of networks at one level of collective action is influenced by that of another level of collective action, and the other way around in recursive ways [1, 21–23]. Such dynamics can be considered to be the outcome of a meta-process bringing together both individuals and organizations, in which the evolution of one level explains in part (in causal terms) the evolution of the other. Level 1 relationships can emerge as a result of the emergence of level 2 relationships. Actors of level 1 may be able under certain circumstances to change the structure of level 2, especially by bringing MLRIs into the picture. MLRIs represent at the same time levers of institutional entrepreneurship and the locus of co-constitution between levels. This is where the two superposed systems of collective action co-evolve and adjust.

As indicated, new families of models are needed to account for such dynamics. One family of models could be a multilevel extension of Snijders [24] model of dynamics of networks, using characteristics of level 2 network as set of exogenous factors in the evolution of level 1 network, and the other way around. At each step of the description of these dynamics at one focal level, information from other, lower or higher, levels must be integrated in the model. The co-evolution of both level networks is "added" to the co-evolution of behavior and relational choices. In terms of model specification, new "independent" variables from inter-organizational networks operate at the inter-individual level, and vice versa. It is perhaps also worth extending Snijders' multilevel version of the model of network dynamics, for example by introducing dual alters or induced potentials, i.e., extended opportunity structures [25], into this controlled formalism. A problem of "synchronization" between levels [26] also emerges. Synchronization is a task of scheduling and coordinating superimposed interpersonal and inter-organizational forms of collective agency, over time and at the cost of one of the levels. Social sciences are currently struggling to measure and model such synchronizations of time scales (short term, long term), especially in political processes where their manipulations can constitute an important competitive advantage.

AMN and MLRIs, especially when they are dynamic, will help better understand politics and multilevel governance in the organizational society, in which superposed levels of collective agency operate, each following their own logic of coordination, while each level is also part of the context of the other levels. This is not trivial since these different logics can be, for example, bureaucratic vs collegial [27–29].

3 Bottom-Up Collegiality, Top-Down Collegiality, and Inside-Out Collegiality

Indeed, organization sociology always starts from an analysis of work, understood in a broad sense as either routine or innovative. From this perspective, each level can be characterized as either predominantly bureaucratic or predominantly collegial [28]. In this dual logics approach, the bureaucratic model is meant to organize collective routine work, concentrate power, command and control unobtrusively at the top, and depersonalize interactions among members. The collegial model is meant to perform collective innovative work with uncertain, unpredictable output and help rival peers self-govern by trying to build agreements and by using private, personalized relational infrastructures to enforce these agreements. Because there are always non-routine tasks to be performed, including that of normative choices and institutional entrepreneurship, organizations are redefined as necessarily combining the two idealtypes for social discipline and productive efficiency, each with its formal and informal dimensions[1]. Both models of organization are needed together in communities, workplaces, markets, and society, and their articulation is under-theorized and under-studied. These combinations of idealtypes must be reassessed in terms of their articulation in real life companies, associations, cooperatives, public authorities, etc. where they are perceived as legitimate or where their legitimacy is contested.

A stratigraphic approach to organized settings [28] shows that "collegial pockets" as social niches capable of collective agency survive in dominant bureaucracies, although with very different and unequal levels of power in regulatory struggles: for example executive suites, professional departments, and workers' trade unions. These collegial levels survive and operate in large bureaucratized and complex organizations when their members are able to come together and learn to defend their regulatory interests. Here AMN shows how the stratigraphic meeting of bureaucracy and collegiality in what we call bottom-up collegiality and top-down collegiality can use MLRIs to strengthen or undermine social participation in organized collective action.

After two centuries of bureaucratization, collegiality as a generic form of organization is really a bottom-up type of collegiality [27] in which collegial pockets of peers—always characterized by oppositional solidarity challenging for incumbent rulers—try to build multilevel relational infrastructures and a presence in the levels of the bureaucracy in which decisive regulation takes place, including the executive level. In the predominantly bureaucratized contexts of contemporary societies, collegiality—where it still exists—is therefore more or less managerialized. Observations of how both models can complement and co-constitute each other in the sense that they drive each other's evolution are provided in recent research. For example, focus on bureaucratic rotation of peers, a process that helps bureaucracies achieve stability from internal movement, provides a first empirical illustration of this dynamic combination. The case of a corporate law firm rotating associates among partners to achieve a balance of powers between rainmakers and schedulers struggling to regulate the organization illustrates this form of combination [10].

Because this combination of logics takes place in an already bureaucratized society, bottom-up collegiality, for example of professionals or trade union members, is often reshaped, and often neutralized, by the bureaucratic ruler, who transforms it into top-down collegiality [27]. The latter is a form of patronage characterized by collegial oligarchies composed by the ruler on a clientelistic basis. Top-down collegiality applies Selznick's [16] cooptation bringing stakeholders into policy-making bodies, but forcing them to turn to this ruler (and to no one else) for help. MLRIs are thus often used by top-down collegiality and AMN is currently being used to provide efficient tools for studying them (for example [28, 29]).

---

[1]To avoid a frequent misunderstanding it is important to stress that collegiality is not the informal dimension of bureaucracy but the organizational idealtype orthogonal to that of bureaucracy. Both bureaucracy and collegiality have their own formal and informal dimensions, their own strengths and weaknesses or vicious cycles.

In such dynamics of multilevel forms of organized collective agency, one particular and contemporary technological evolution deserves special attention for its social implications. One of the most phenomenal contemporary innovations is the digitalization of interactional and relational life with online social networks. In our view, these online networks boost the bureaucratic systematic capacity to monitor, reshape, and routinize collegial pockets, the very core makeup of collegiality: personalized relational activity and MLRIs. We call "inside-out collegiality" the combination of the two logics in which bureaucratic digital framing, parametrizing, monitoring and control of private personal relationships (made transparent to owners of the platform) shape collective agency in order to strip collegiality of its oppositional solidarity. In that sense, inside-out collegiality not only strengthens neo-liberal individualization and flexibilization of labor markets but threatens institutional entrepreneurship and the political process as defined above.

The struggle and co-constitution between the two idealtypes thus takes a dramatic turn. Digitalization as contemporary bureaucratization turns the bottom-up collegial model "inside out," deepening bureaucratization of collective action and society [30]. Freedoms and privacy, oppositional solidarities, and capacity to innovate are deeply threatened by what amounts to using organizations as tools for imposing new forms of collective responsibility and for further dividing societies between the many and the few [31]. Struggles to find new forms of collegiality in cooperatives, in the commons and in more distributed uses of platforms, such as new peer-to-peer innovations, resist such developments and would benefit from better knowledge of dynamics of multilevel networks in new forms of organized collective agency.

4 An Example of Top-Down Collegiality in Institutional Entrepreneurship

An example can be provided in a study of the emergence of a new European intellectual property regime via the construction of a transnational court, the European Unified Patent Court (UPC) [32]. This court is considered by European industries with patents at the core of their business model as important to strengthening a contemporary European knowledge economy, including promotion and protection of innovation. The construction of this institution requires institutional entrepreneurship involving individuals (professionals), organizations, and governments. "Harmonization" of a variety of national legal frameworks has required MLRIs for coordination between networks of individuals, networks of organizations, and cross-level coordination between networks of individuals and organizations. Neither individuals, nor organizations, nor governments could access or mobilize, on their own and at the right time, all the resources that are needed to be efficient in this institutionalization process. Structuration at one level drove structuration at the other, often in conflicting and unequal ways. Time to adjust and adapt was available to some, but not to others in dynamic and multilevel political construction.

Top-down collegiality accounts well for the construction of the UPC. With help from Brussels bureaucrats and from a professional association of corporate lawyers, a powerful, public-private European agency, the European Patent Office (EPO), sole regulator of intellectual property at the European level in the absence of a transnational court, a collegial oligarchy of national judges specialized in patents was selected as patent experts and assembled at the so-called Venice Forum, a private field-configuring event. Based on this top-down cooptation, a core group among these judges was then promoted as an ex ante leadership into a collegial oligarchy that was able to define the Rules of Procedure of the future UPC. They were punching above their weight in the regulatory process of harmonization of divergent national legal frameworks into a single body of rules under which the future institution would operate. The bureaucratic ruler in Brussels allied with EPO operated top-down through a form of patronage, selecting judges with strong multilevel status or promoting others to this status. This top down selection of a collegial oligarchy of ex ante leaders was instrumental for the development of the project, neutralizing in particular civil society actors opposed to the ways in which patents are used in contemporary capitalism, i.e. as financial instruments paradoxically undermining open science and increasingly innovation itself. Contemporary institutions are increasingly designed, operated and evaluated by such top down collegial oligarchies.

One of the problems for such politics is precisely a problem of coordination of the regulatory processes that occur at one level with the same processes occurring at the other levels, i.e., "harmonization" of different time frames, sources of normativity and governance within and across levels. How this takes place is still not very well known in detail and can be investigated with AMN. Today, there are no tools for evaluating the vast dynamic and multilevel worldwide rule-making activity in any comprehensive way. New institutions arise when organized actors with sufficient resources see in them an opportunity to realize interests that they value highly [33, 34]. These "institutional entrepreneurs" struggle over which institutional arrangements to select for the collective. MLRIs and vertical linchpins driven by top-down collegiality in superposed levels of collective agency are thus key to policy- and rule-making is all domains of life: water management, food, health and safety, transportation, etc. An unknown number of discreet collegial oligarchies acquire the right kind of structural, cross-level position in such multilevel governance systems and create regulatory regimes that are not accountable to the public.

In particular, institutional entrepreneurship requires a global vision of this multilevel system. Actors at different levels do not have the same resources and capacities to build this vision and to promote and protect their regulatory interests. In regulatory competition between strata of collective agency (local, national, international), the issue of how formal and informal knowledge networks and rule-making behavior influence each other converge or diverge in terms of building institutions that will be considered to be legitimate, this issue is thus a crucial problem of dynamics of multilevel networks. The latter co-evolve with normative action taking place in several superposed political arenas, whether public, private (closed), or a mix of public/private, and are very complex to grasp. Usually, transnational private regulation that has been spreading globally pretends that it solves the problem of this competition between levels and stakeholders by providing flexible guidelines, a general normative baseline that is adaptable to local situations via subsidiarity, thus helping each level protect its regulatory interests as it sees fit. However these rhetorics are part of the process and need to be factored into the analyses as well, thus requiring dynamics of multilevel networks to combine structure, culture, and agency.

5 The Challenge of Contextualizing Multilevel Networks: Organized Mobility and Relational Turnover

Building and maintaining MLRIs and institutions is not a collective adventure that takes place in a vacuum, but in Selznick's [16] dynamic conditioning fields. To understand MLRIs and their role in synchronization of levels of collective action, it is useful to see them as determined in part by organizational mobility of members at each level and by subsequent relational turnover in their respective networks (OMRT). The word "organized" is used to qualify mobility because both social actors and the social system create paths and rules for movements and careers (for incoming, rotating, reshuffled, promoted, demoted, outgoing actors) that are not allowed to be random [35, 36]. Multilevel positioning can be complex because mobility in turn produces relational turnover for these members and this turnover is managed by the creation of the new relational infrastructures, for example specific forms of multilevel social status. Efforts to synchronize the temporalities of the levels create the energy for more intra- and inter-organizational mobility and controversies. Synchronization costs must then include efforts spent to position oneself in the dynamic conditioning fields at the different levels of social space so as to be able to build or maintain MLRIs. Incurring synchronization costs will be rewarding (in terms of managing constraints, learning, making one's voice heard in controversies, and regulation) for some players; for others, who are unable to capitalize on social resources thanks to the maintenance of such multilevel relational infrastructures, they will amount to sunk costs.

Actors can experience OMRT as new contextual constraints and opportunities, especially as possible emancipation from constraints imposed by prior affiliations, or as networks to nowhere, or as opportunities to introduce organizational change by bringing in new members. To some extent, institutional entrepreneurs attempt to use OMRT to reshape this multilevel structure—often with unexpected consequences. Such dynamics are not visible enough, for example, in current studies of social inequalities. A dynamic and multilevel network approach to social life changes the measurements of these socio-economic costs precisely by introducing more complex and systematic positioning, mobility, and relational turnover into the picture of management of inequalities.

This assumes that some uses of MLRIs such as multilevel social niches and multilevel status (for example vertical linchpinship) are both building blocks for cross-level synchronization and instruments of restructuration attempts across levels. The connection between mobility and relational turnover is often explored in part and in depth in specific areas of social life. Often overlooked in the literature are the general effects of this systematic, recursive, and transformative link between the two realities (mobility across systems of places and relational capital) and its implications for social life. There are connections between these movements, as actors switch places in these circuits, and change—at least in part— their normative choices and respective sets of relationships, i.e., their respective relational capital. There is also an effect of the latter changes on the evolution of the system of places itself, an evolution that is only visible if places are not considered as purely contextual and exogenous, but as accumulated by actors—and thus as endogenous in the mechanisms under examination and models that account for them. Combining mobility in loops [35] and co-evolution of multilevel networks and behavior [37] helps make institutional entrepreneurship and OMRT structuration, with their multilevel dynamics and associated synchronization costs, measurable, and more generally redefine the social costs of living in an organizational and market society.

6   Multispin for Contextualizing Multilevel Networks

Whether physical (for example through migration) or social or both, these articulated movements and changes represent important determinants of social structure, order, and inequalities in the organizational society. They are created by the social organization of these *milieux* and end up, under conditions that remain to be spelled out, restructuring these *milieux*, promoting some members in terms of ability to define new norms, and pushing others out of the regulatory process.

This is where an overall theoretical link is needed between OMRT as forms of contextualization of networks, MLRIs, and organizational analysis of collective action. We propose, as an initial step, a guiding metaphor for this link in the picture of a multilevel spinning-top, or "multispin" (see Fig. 2). This metaphor is too rigid for many purposes, but helpful nevertheless as an initial heuristic for representation of the dynamic conditioning fields of institutionalization processes [36] because it is a dynamic structure combining several sub-processes in which movement creates stability, thereby promoting some actors and expelling others as in musical chairs. In our view, this image of a rotating three-level structure provides intuitions for contextualizing the emergence of institutions as a dynamic multilevel process. It helps explain how a small collegial oligarchy of networked institutional entrepreneurs with multiple and inconsistent forms of status [10] uses, in its lobbying activity, multilevel position in these networks and their dynamics. Stability from movement in the multispin helps institutional entrepreneurs acquire the staying capacity and subsequent influence that is needed to frame, build, and entrench new institutions.

Fig. 2 Multilevel
spinning-top with staircase in
the shaft, or multispin, a
metaphor for organized
mobility and relational
turnover. Design: Elie
Partouche

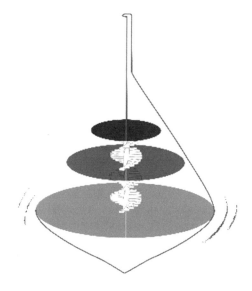

In this metaphor, each level represents a network of collective action. In the emergence of institutions, the bottom level is composed of citizens, the second level of private organizations and public institutions, and the third level as governments, national and transnational. Affiliations as links between levels are not displayed in the picture for the sake of lisibility. Analytically speaking, agency starts at the level of individual networks. The evolution of these networks—each at its own level but influencing the evolution at the other levels through synchronization— is driven in part by controversies and mobility of actors moving into this system from the outside. The core set of individual institutional entrepreneurs with super-central status moves up the shaft and acquires a competitive advantage in the joint regulatory process of institutionalization. Transferring synchronization costs is rewarding for these actors when they have a strong multilevel position because these costs are either shared or dumped on others, who can end up in the periphery or in limbo. For example, revolving doors from public responsibilities to private jobs and back to public positions help create this informal pecking order and concentrate power with help of conflicts of interests. In this example, it is not enough for institutional entrepreneurs to have an official mandate to build an institution. Superposition of these dynamic relational systems of collective action and coordinated activities between them must provide these super-central entrepreneurs with sufficient resources, staying capacity, stability, and legitimacy to drive the institution-building process over time, long enough for the institution to emerge and/or change.

Multispin is a first metaphor meant to contextualize MLRIs and social processes that individual and collective actors navigate in Selznick's dynamic conditioning (mine) fields. More generally, this metaphor accounts for the systematic rotation— such as job rotation—from one place to another in a system of places, a movement that creates relational turnover in members' personal networks. Over time, this relational turnover tends to slow down because members manage turnover by turning to a small and stable set of authoritative contacts, for example super-central advisors, who can then be compared to members who climbed the stationary shaft of the multispin, a metaphor for social status as MLRI represented as a staircase. Indeed members can gain or lose multilevel status and vertical linchpinship, i.e., capacity to act at different levels simultaneously, just like stairs can lead up or down. They can rise upwards, usually to dominate, or sink downwards, usually to be pushed out of the regulatory process. They then gain or lose influence as institutional entrepreneurs because the tendency to turn to a small and stable set of authoritative contacts creates a central core at the next level higher up, ultimately becoming ratchets of social stratification [37]. In short, this metaphor brings together individual and collective actors, trajectories, relational turnover in actors' networks, actors' multilevel status measured by centrality in superposed, overlapping networks, decisions, and normative choices. This structure however can also lose its balance and the process fail, unless all these ingredients [38] are kept together by the energy coming from socially organized mobility, for example resilience from MLRIs.

Multispin accounts for this institutionalization process in the empirical example presented above. Judges were brought to the Venice Forum on a top-down collegiality basis, then circulated across Europe to learn from each other and identify the ex ante leadership that was promoted to the collegial oligarchy with enough staying capacity, at least at two levels simultaneously, to become the permanent interlocutors of Brussels and EPO (the top level). Expert personnel was also circulated between corporate law firms, national ministries of justice, law schools in universities, courthouses, training facilities for future European judges, and industry associations, accounting for rotation at the medium inter-organizational level. Rotation at the level of governments was perhaps slower, less fluid than expected by the business communities bringing together large corporations, slowing down the process, and increasing synchronization costs for the levels below, to the point that the institutionalization process stalled, as if the multispin had stopped and fallen down.

7   Conclusion

This paper argues that the analysis of multilevel networks is useful to understand politics, institutional entrepreneurship, and social change. Investigating these fundamental realities and phenomena requires combining inter-individual networks and inter-organizational networks of institutional entrepreneurship over time. AMN helps identify multilevel relational infrastructures (in particular multilevel social status) on which institutional entrepreneurship depends, especially in collegial oligarchies as laboratories for social change. In heavily bureaucratized societies, these laboratories can take various forms such as bottom-up collegiality, top-down collegiality, and inside-out collegiality. We argue that, in an era of vital transitions, one of the main challenges for social network analyses is to use AMN to observe these collegial oligarchies and to model and understand social (in)capacities to build alternative multilevel relational infrastructures promoting social change. This challenge leads to another: that of understanding the conditions under which a form of collegiality is selected by contextualizing institutional entrepreneurship and its multilevel relational infrastructures. The paper theorizes organized mobility and relational turnover as important dimensions of this contextualization of institutionalization processes.

These analyses have the potential to play an important role in society, when faced with transitions-related challenges. In contemporary organizational societies, giant private companies create collegial oligarchies by using their privatized multilevel network data and instruments of inside-out collegiality for private institution building with questionable legitimacy: for example by reshaping entire cities with apparent democratization of new technologies of decentralization of services (blockchains); by developing private community self-organization with parametrized digital platforms for management of local resources, often competing with the public political architecture of these communities; by creating private currencies; by monopolizing relational data and building low quality social sciences (undermining high quality open science) for brute force social engineering. Studying Selznick's "dynamic conditioning fields" as OMRT contextualizing forms of collegiality and institutional entrepreneurship might help understand these processes so as to keep multilevel political steering of future development accountable and democratic.

References

1. Lazega, E., Jourda, M.-T., Mounier, L., Stofer, R.: Catching up with big fish in the big pond? Multi-level network analysis through linked design. Soc. Networks. **30**, 159–176 (2008)
2. Breiger, R.L.: The duality of persons and groups. Soc. Forces. **53**, 181–190 (1974)
3. Fararo, T.J., Doreian, P.: Tripartite structural analysis: generalizing the Breiger Wilson formalism. Soc. Networks. **6**, 141–175 (1980)
4. Wasserman, S., Iacobucci, D.: Statistical modelling of one-mode and two-mode networks: simultaneous analysis of graphs and bipartite graphs. Br. J. Math. Stat. Psychol. **44**, 13–43 (1991)
5. Wang, P., Robins, G., Pattison, P., Lazega, E.: Exponential random graph models for multilevel networks. Soc. Networks. **35**, 96–115 (2013)
6. Perrow, C.: A society of organizations. Theory Soc. **20**, 725–762 (1991)
7. Thelen, K.: How Institutions Evolve: The Political Economy of Skills in Germany, Britain, the United States and Japan. Cambridge University Press, Cambridge (2004)
8. Hoffman, A.J.: Institutional evolution and change: environmentalism and the US chemical industry. Acad. Manag. J. **42**, 351–371 (1999)
9. Meyer, J., Scott, R. (eds.): Organizational Environments: Ritual and Rationality. Sage, Beverly Hills, CA (1983)
10. Lazega, E.: The Collegial Phenomenon: the Social Mechanisms of Cooperation among Peers in a Corporate Law Partnership. Oxford University Press, Oxford (2001)

11. Lazega, E.: Networks and institutionalization: a neo-structural approach. Connections. **37**, 7–22 (2018)
12. Glückler, J., Suddaby, R., Lenz, R. (eds.): Knowledge and Institutions, Knowledge and Space Series, vol. 12. Springer, Dordrecht (2018)
13. Weber, M.: Economy and Society. University of California Press, Berkeley (1920/1978)
14. Olson, M.: The Logic of Collective Action. Harvard University Press, Cambridge (1965)
15. Wittek, R., Van de Bunt, G.G.: Post-bureaucratic governance, informal networks and oppositional solidarity in organizations. Neth. J. Soc. Sci. **40**(3), 295–319 (2004)
16. Selznick, P.: TVA and the Grassroots: a Study in the Sociology of Formal Organization. University of California Press, Berkeley (1949)
17. Snijders, T.A.B.: The multiple flavours of multilevel issues for networks. In: Lazega, E., Snijders, T.A.B. (eds.) Multilevel Network Analysis for the Social Sciences; Theory, Methods and Applications, pp. 15–46. Springer (2016)
18. Žiberna, A.: Blockmodeling of multilevel networks. Soc. Networks. **39**, 46–61 (2014)
19. Barbillon, P., Donnet, S., Lazega, E., Bar-Hen, A.: Stochastic block models for multiplex networks: an application to a multilevel network of researchers. J. R. Stat. Soc. A. Stat. Soc. **180**(1), 295–314 (2017)
20. Bar-Hen, A., Barbillon, P., Donnet, S.: Block models for multipartite networks. Applications in ecology and ethnobiology. arXiv preprint arXiv:1807.10138 (2018)
21. Berends, H., Van Burg, E., van Raaij, E.M.: Contacts and contracts: cross-level network dynamics in the development of an aircraft material. Organ. Sci. **22**, 940–960 (2011)
22. Berge, C.: Hypergraphs: Combinatorics of Finite Sets, vol. 45. Elsevier, Amsterdam (1984)
23. Grossetti, M.: L'espace à trois dimensions des phénomènes sociaux. Echelles d'action et d'analyse, SociologieS. http://sociologies.revues.org/index3466.html (2011)
24. Snijders, T.A.B.: Stochastic actor-oriented models for network change. J. Math. Sociol. **21**, 149–172 (1996)
25. Lazega, E., Jourda, M.-T., Mounier, L.: Network lift from dual alters: extended opportunity structures from a multilevel and structural perspective. Eur. Sociol. Rev. **29**, 1226–1238 (2013)
26. Lazega, E.: Synchronization costs in the organizational society: intermediary relational infrastructures in the dynamics of multilevel networks. In: Lazega, E., Snijders, T. (eds.) Multilevel Network Analysis: Theory, Methods and Applications. Springer, Dordrecht (2016)
27. Lazega, E., Wattebled, O.: Two definitions of collegiality and their inter-relation : The case of a Roman Catholic diocese. Sociol. Trav. **53**(Supplement 1), e57–e77 (2011)
28. Lazega, E.: Bureaucracy, Collegiality and Social Change: Redefining Organization with Multilevel Relational Infrastructures. Edward Elgar, Cheltenham (2019)
29. Richard, C., Wang, P., Lazega, E.: A Multilevel Network Approach to Institutional Entrepreneurship: the Case of French Public-Private Partnerships (2019)
30. Lazega, E.: Networks and commons: bureaucracy, collegiality and organizational morphogenesis in the struggles to shape collective responsibility in new sharing institutions. In: Archer's, M.S. (ed.) Morphogenesis and Human Flourishing, vol. V, pp. 211–237. Springer, Dordrecht (2017)
31. Lazega, E.: Swarm-teams with digital exoskeleton: On new military templates for the organizational society. In: Al-Amoudi, I., Lazega, E. (eds.) Post-Human Institutions and Organizations: Confronting the Matrix. Palgrave, Basingstoke (2019)
32. Lazega, E., Quintane, E., Casenaz, S.: Collegial oligarchy and networks of normative alignments in transnational institution building: the case of the European Unified Patent Court. Soc. Networks. **48**, 10–22 (2016)
33. Eisenstadt, S.N.: Cultural orientations, institutional entrepreneurs, and social change: comparative analysis of traditional civilizations. Am. J. Sociol. **85**(4), 840–869 (1980)
34. DiMaggio, P.: Interest and agency in institutional theory. In: Zucker, L. (ed.) Institutional Patterns and Organizations: Culture and Environments, pp. 3–21. Ballinger, Cambridge, MA (1988)
35. White, H.C.: Chains of Opportunity: System Models of Mobility in Organizations. Harvard University Press, Cambridge, MA (1970)

36. Lazega, E.: Organized mobility and relational turnover as context for social mechanisms: a dynamic invariant at the heart of stability from movement. In: Glückler, J., Lazega, E., Hammer, I. (eds.) Knowledge and Networks, Volume 11 of the Series Knowledge and Space, pp. 119–142. Springer, Dordrecht (2017)
37. Tilly, C.: Durable Inequality. University of California Press, Berkeley (1998)
38. Lazega, E.: Joint 'anormative' regulation from status inconsistency: A multilevel spinning top model of specialized institutionalization. In: Archer, M.S. (ed.) Anormative Regulation in the Morphogenic Society, pp. 169–190. Springer, Cham (2016)

Part I
Methods

Socio-Cultural Cognitive Mapping to Identify Communities and Latent Networks

Iain Cruickshank and Kathleen M. Carley

Abstract Deriving networks and communities from individual and group attributes is an important task in understanding social groups and relations. In this work we propose a novel methodology to derive networks and communities from socio-cultural data. Our methodology is based on socio-cultural cognitive mapping (SCM) and k-NN network modularity maximization (SCM + k-NN) that produces both a latent network and community assignments of entities based upon their socio-cultural and behavioral attributes. We apply this methodology to two real-world data sets and compare the community assignments by our methodology to those communities found by k-Means, Gaussian Mixture Models, and Affinity Propagation. We then analyze the latent networks that are created by SCM + k-NN to derive novel insight into the nature of the communities. The community assignments found by SCM + k-NN are comparable to those produced by current unsupervised machine learning techniques. Additionally, in contrast to current unsupervised machine learning techniques SCM + k-NN also produces a latent network that gives additional insight into community relationships.

Keywords Social network analysis · Community detection · Clustering

1 Introduction

An important task in social science research is to understand communities within a population and their relationship dynamics. Furthermore, these communities and relationships in the population are not known explicitly; we only have collected socio-cultural and behavioral attributes and must infer what communities exist and how they are related. In this work, we build upon previous work in socio-cultural cognitive maps (SCMs) to find these latent communities and network structures

I. Cruickshank (✉) · K. M. Carley
Center for Computational Analysis of Social and Organizational Systems (CASOS), Institute for Software Research, Carnegie Mellon University, Pittsburgh, PA, USA
e-mail: icruicks@andrew.cmu.edu; kathleen.carley@cs.cmu.edu

© Springer Nature Switzerland AG 2020   35
G. Ragozini, M. P. Vitale (eds.), *Challenges in Social Network Research*,
Lecture Notes in Social Networks, https://doi.org/10.1007/978-3-030-31463-7_3

based upon recorded socio-cultural and behavioral attributes. Our approach provides a novel way of fusing together various socio-cultural and behavioral attributes into a network model that can be used with well-founded network analysis techniques to derive insight about real-world communities.

In this work, we extend the SCM framework to find a latent graph and communities. The SCM framework finds the best fit coordinates in a N-dimensional space for a set of entities based upon an attenuated Minkowski distance applied to their observed attributes [1, 2]. With the coordinates produced by the SCM procedure, we can then measure the distance between points, such that those points which are closer are more similar in their socio-cultural attributes. Using that distance, we construct k-NN networks by considering the k nearest neighbors of each point. We then select those latent networks which produce the best modularity value for a modularity subgrouping routine [3]. The final result is both the latent network and community assignments for all of the socio-cultural data. Using this method, we apply it to two different real-world data sets and analyze the resultant latent communities and networks to demonstrate the utility of both the latent communities and latent networks in analysis of social science data.

2 Related Work

The work presented in this paper falls within the general category of *Latent Space Models*. In general, latent space models typically work by placing entities in some kind of graph based upon the similarity (or distance) based on the observed attributes of the entities [4–6]. This is typically done by positioning nodes within an N-dimensional space and then creating edges between those nodes that are closer together [6]. For nearly all latent space models, the underlying assumption of the models is that nodes within an N-dimensional space have ties between them based upon independent conditional probabilities [6, 7]. Regarding these models, there are generally two ways to estimate the probability distributions for edge formation: a distance model and a projection model [6]. In the distance model, nodes that are closer in a latent space tend to form links. In the projection model, nodes that have a narrower angle (as measured between two vectors extending from the origin to the points) between their spatial vectors tend to form links; a narrower angle implies that the points are closer in the latent space. These probabilistic models have been further extended to dynamic networks and multiplex or multidimensional networks [5, 6]. All of these models rely on the assumption that the formation of links between any given set of nodes in the latent space is statistically independent of any other link forming between any given set of nodes, which allows for fitting the models via Bayesian methods.

When it comes to measuring communities or clusters within a network there are many measurements, such as conductance, expansion, internal density, and many others [8]. One of the most commonly used metrics is modularity [4, 9]. Roughly defined, modularity is the fraction of the links that fall within the defined communities of a network minus the expected fraction if links were distributed at random throughout the network. There have been many successful community detection algorithms for networks based upon the measure of modularity [10]. In this work, we will employ both the measure of modularity, as well as one of the more well-known and fast algorithms for modularity-based community detection, the Louvain Method, as part of the process of finding a network in the latent space.

Another area of research that is closely related with our work is dimensionality reduction. Some classical examples of dimensionality reduction methods are Multidimensional Scaling (MDS) and Principal Component Analysis (PCA) [11]. Both of these techniques suffer from an assumption of linearity in their models. This shortcoming has been subsequently overcome by another common dimensionality reduction technique, t-Stochastic Neighbor Embeddings (t-SNE) [12]. Much like t-SNE, SCM embeds nodes in a particular space of a different number of dimensions than the original data (usually, fewer dimensions), which enables visualization and clustering [2]. As noted in previous work on SCMs, there are some distinct differences between SCM and more standard dimensionality reduction procedures [2]. In particular, SCM does not rely on specification of a distance metric for the latent space, but rather determines one through its optimization procedure. This flexibility allows SCM to vary the amount of impact the position of one node has on position of other nodes when placing nodes into a latent space. Additionally, unlike t-SNE which uses statistics for continuous or binary data to evaluate fit, SCM is custom suited to handle categorical data, which occurs in sociological and behavioral studies.

3   Method

To begin, we will briefly describe the SCM process, but also refer the reader to [1] for the precise details of the formulation of an SCM process and [2] for an implementation of an SCM process. The first step in the SCM process will be to convert the data matrix into a frequency matrix. As an example, a data matrix could be a set of actors with their associated attributes, and at each attribute, an actor can have a series of categorical responses. It should be noted that all attributes are considered equal; there is no distinction between behavioral and socio-cultural attributes. This matrix is then converted to a frequency matrix where the entries are the counts of the shared levels of every attribute between each of the actors. So, the frequency matrix is of size actor-by-actor, and has entries that are the count of the number of same attribute values for all attributes of the data matrix between each actor and every other actor.

Next, we then find the positions in a latent space for each of the data points such that the function determining the distances produces fitted frequencies that are as near as possible to the observed frequencies from the previous step. The fitted frequencies for each data point are calculated by

$$F(i, j) = R_i \times C_j \times 2^{-d_{ij}^a} \tag{1}$$

where R_i is the row factor term, C_j is the column factor term, and $2^{-d_{ij}^a}$ is the row and column correlation factor, where a is an *attenuation* term. a controls how much impact the interaction between two entities has on their co-occurrence frequency. A smaller a results in the distance affecting the frequency count to a smaller extent as the distance between any given data points increases. When $a = 2$ the interaction between row and column variables is modeled as a Gaussian Distribution. The distance metric, d_{ij} is defined as the Minkowski distance:

$$d_{ij} = \left(\sum_{dim=k}^{ndim} |x_{ik} - y_{jk}|^M \right)^{1/M} \tag{2}$$

where M is the *power* setting of the metric that determines the space that the points are mapped into (i.e., $M = 2$ produces an Euclidean distance and space). So, the SCM process then proceeds as an optimization problem where the points of each actor are inferred in a latent space with the target function as the χ^2 value between the fitted frequencies, $F(i, j)$ and the observed frequencies, for user supplied attenuation, power, and number of dimensions. The overall process of the SCM can be summarized as follows (Fig. 1 further illustrates the SCM procedure):

1. Select a set of power and attenuation settings and number of dimensions for the SCM procedure. Typical selections are $a = 0.7, 1, 2, 3$ and $M = 0.7, 1, 2, 3$, and two dimensions for the inferred points of the agents.
2. For each combination of the parameters of a and M, do the following:

   – Place each data point in a latent space where the number of dimensions of the latent space is specified by the user, and calculate the frequencies of this placement using Eqs. 2 and 1.

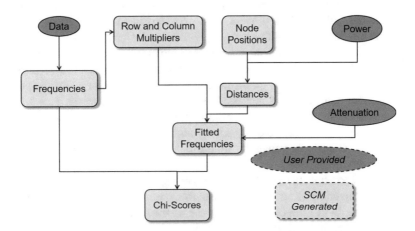

Fig. 1 Flowchart of the SCM process. User inputs are the number of dimensions, power, and attenuation, and outputs are the latent points of the nodes and the distances between the nodes in the latent space

– Evaluate the current placement using the result of the last step and observed frequencies using a χ^2 value.
– Continue to adjust the placements of the points to achieve a better χ^2 value.

3. Return the coordinates of those placements of the points that have the best χ^2 value.

Once we have obtained the pairwise distances between each of the actors, we can then use k-NN network modularity maximization to find the latent network and subgroup assignments of the actors. We will leave the technical details of the algorithm to [3] and [13], but will give a brief description of the algorithm, with our changes to it. k-NN network modularity maximization takes an affinity or distance matrix and creates a network where each node connects to its k nearest neighbors. Then, this network is clustered using the Louvain method of modularity maximization [14]. In clustering step, we differ from the original algorithm proposed in [3], as we use a faster method of modularity maximization of unimodal networks, the Louvain Method, as opposed to the author's *QCut* algorithm. For our methodology, we are using an asymmetric k-NN network, where two nodes are connected if one or both of the nodes feature the other node in their k nearest neighbors [13]. More precisely, for each point x_i, let $N_k(x_i)$ be the k nearest neighbors of x_i, then an asymmetric k-NN network has links between two points x_i and x_j if $x_i \in N_k(x_j)$ OR $x_j \in N_k(x_i)$. The pseudocode of our implementation of the k-NN modularity maximization procedure is detailed in Algorithm 1.

Algorithm 1 k-NN modularity maximization procedure

---

input: Distance or Affinity Matrix, S, (n x n)
for $i = 1 : \lfloor log_2 n \rfloor$ **do**
   $k \leftarrow 2^i$
   $G_k \leftarrow kNN(S, k)$
   $C(G_k) \leftarrow Louvain(G_k)$
   $G_k^r \leftarrow randomize(G_k)$
   $C(G_k^r) \leftarrow Louvain(G_k^r)$
   $Modularity_k \leftarrow Modularity(C(G_k)) - Modularity(C(G_k^r))$
end for
$k^* \leftarrow argmax_k Modularity_k$
$G^* \leftarrow kNN(s, k^*)$
$C(G^*) \leftarrow Louvain(G^*)$
return $C(G^*), G^*$

With these latent networks and subgroups, we can then address our initial questions regarding possible groups and network structures in a community based on measured attributes of the community. To summarize, the SCM plus k-NN modularity maximization procedure works as follows:

1. Select a set of power and attenuation settings and number of dimensions for the SCM procedure. Typical selections are $a = 0.7, 1, 2, 3$ and $M = 0.7, 1, 2, 3$, and two dimensions for the inferred points of the agents.

2. Run the SCM procedure for each of the attenuation and power settings combinations. Select the output from those settings that produce the minimal χ^2 value. Find the distance between each point and every other point based upon the coordinates found by the SCM procedure, and store these in a distance matrix.
3. Using the distance matrix, perform k-NN network modularity maximization to find the optimal subgroups and latent network.
4. Analyze the subgroups and latent networks using standard network analysis techniques to answer research questions.

4   Empirical Results

To evaluate the SCM plus k-NN network modularity maximization procedure (SCM + k-NN), we will consider two real-world data sets and compare the subgroups obtained to those obtained by other unsupervised techniques. Since we are interested in methods that also determine the number of subgroups without a priori input, we will consider only other unsupervised machine learning techniques that do not require specification of the number of subgroups. Namely, we will use k-Means, where k is determined by Bayesian Information Criteria (BIC) [15], a Gaussian Mixture Model where the number of Gaussians is determined BIC [16], and affinity propagation [17]. For each of the experiments, we used the program Organizational Risk Analyzer (ORA) version 3.0.9 to perform the SCM with settings of $a = 0.7, 1, 2, 3$ and $M = 0.7, 1, 2, 3$ with two dimensions for the inferred points in the latent space [18]. All visualizations and standard network analysis techniques were also performed in ORA. For the other three methods, similarity was determined using Euclidean distance.

Since these data sets do not have a ground truth subgrouping, we will analyze the subgroups for qualitative meaning and compare different subgrouping results using the Adjusted Rand Index (ARI) [19]. ARI is a measure of the similarity of two clusterings; it gives an idea of how often two different methods will assign the same labels to the same sets of data. We will also use Silhouette score to measure the performance of each of the clustering techniques on each of the data sets [20]. The two data sets are as follows:

- **Hatfield McCoy Data Set**: Data set is a set of actors with their associated attributes from the famous historical feud between the Hatfield and McCoy families from 1863 to 1891 [2, 21]. The data set was compiled from various historical records and includes individuals attributes from the time of the feud (1863–1891). It consists of 66 individuals with 9 attributes that are all binary valued (e.g., member of Hatfield family, Female, etc.)
- **8th Convocation of the Ukrainian Parliament**: Data set consists of the set of the parliamentarians of the 8th (current) convocation of the Ukrainian Parliament with some of their publicly available attributes and their votes on the bills that were proposed by the Ukrainian president [22]. The data set has 522

parliamentarians with their votes on 62 different presidential bills. Vote responses can take one of five values: yea, nay, abstain, absent, did not vote (some members of the parliament were not members of the parliament for all of the convocations' votes). The data set also contains 10 categorical and real-valued attributes, such as age, former professions, etc.

4.1 Hatfield–McCoy Case Study

Beginning with the Hatfield McCoy data set, the optimal SCM parameters were $a = 2$, $M = 0.7$, and a goodness-of-fit of $\chi^2 = 32.36$. Table 1 summarizes the summary statistics of the different subgroups identified in the data by the different methodologies:

Affinity propagation and SCM + k-NN both seem to have produced very similar subgroups, with a larger number of subgroups relative to both GMM and k-Means. Furthermore, both affinity propagation and SCM + k-NN produced higher values of the silhouette which indicates less overlap between subgroups than those subgroups found by GMM or k-Means. Looking at the membership of the different subgroups, k-Means seems to have almost entirely split upon whether the person was a Hatfield or a McCoy; the two subgroups are the two families. The subgroups produced by the other three methods have much more nuance. For example, SCM + k-NN has separate subgroups for the "Devil Anse Kids" and the "Randolph Kids" which are both subsets of the Hatfield's and the McCoy's, respectively.

We next analyze the subgroups found by the various methods by ARI, as depicted in Fig. 2. As noticed in the summary statistics the subgroup assignments of individuals people are highest between SCM + k-NN and affinity propagation. k-Means had subgroup assignments least like any other subgrouping method. Therefore, it would seem SCM + k-NN produced meaningful and more nuanced subgroups comparable with affinity propagation. One distinct advantage of SCM + k-NN modularity, however, is that we also obtain a latent network, which is visualized in Fig. 3.

In the resulting network, two of the most interesting groups of individuals, the "Devil Anse Kids" and the "Randolph Kids," are placed into completely separate components. Additionally, running standard unimodal network centrality measures produces some interesting results. Ephraim Hatfield and America McCoy, both of

Table 1 Summary statistics of subgroups found in Hatfield McCoy data set

Subgrouping method	Number of subgroups	Avg. subgroup size	Std. subgroup size	Range of subgroup sizes	Silhouette
SCM+k-NN	11	6.0	1.89	[3,10]	0.411
GMM with BIC	4	16.5	6.61	[12,26]	0.319
k-Means with BIC	2	33.0	4.24	[30,36]	0.378
Affinity propagation	11	6.0	3.16	[3,11]	0.691

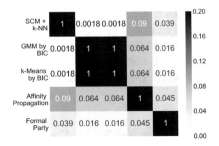

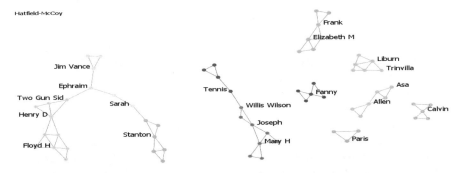

(a) ARI values between different sub grouping methods. Affinity Propagation and SCM + k-NN have the most similar sub group assignments for each of the individuals.

(b) ARI values between different sub grouping methods. Both GMM and k-Means produced exactly the same sub group assignments, while no sub grouping method was close to the same assignments as the formal parties of the parliamentarians.

Fig. 2 ARI plots of subgroups for (**a**) Hatfield McCoy and (**b**) Ukrainian Parliament

Fig. 3 Latent network of the Hatfield McCoy data set. Coloring represents subgroup assignments. The low, uniform values of degree for the nodes indicate that actors are spread out in their socio-cultural attributes than sharing many of the same socio-cultural attributes. Areas of high socio-cultural similarity would more likely result in a more hub-and-spoke style of network

whom were intermarried to the members of the other family, ranked highest in betweenness centrality, which likely reflects the bridging roles they played by being intermarried. Floyd Hatfield, Henry Hatfield, and Mary Hatfield have the highest degree centrality values in the latent graph, which indicates the Hatfields have more central and shared socio-cultural attributes relative to the entire population than the McCoys. Furthermore, nearly all of the Hatfield members are within the same component, whereas the McCoy members are split more evenly across multiple components of the latent graph. This property of the latent graph may indicate that the Hatfields possess more socio-cultural similarity than the McCoys, which could have impacted the historical course of events of the Hatfield McCoy feud. Additionally, while there are two main components in the network, these components are not densely interconnected, possibly indicating that the socio-cultural latent space is more spread out and homogenous and there are no particularly dense regions of socio-cultural attribute values. If each actor were allowed to increase its number of links, it is likely the modularity of the overall graph and subgroups would decrease.

4.2 Ukrainian Parliament Case Study

Turning now to the second data set of the 8th Convocation of the Ukrainian Parliament, the optimal SCM parameters were $a = 2$, $M = 2.0$, and a goodness-of-fit of $\chi^2 = 88,896.69$. Table 2 reports the summary statistics of the different subgroups identified in the data.

As with the previous data set, we find first that the grouping statistics between affinity propagation and SCM + k-NN are more alike and the grouping statistics between k-Means and GMM are more alike. These differences are particularly pronounced in the sizes and number of the subgroups found in the data. This difference in size and number of groups is likely also a contributor to the relative differences in the silhouette score between the two sets of methods. The generally lower values of the silhouette score for this data set likely indicate that it is a difficult data set to subgroup, as there is a lot of overlap between the attributes and behavior of parliamentarians, and so finding a smaller number of subgroups has likely led to more overlap between the subgroups. One major difference with the SCM + k-NN procedure on this data set is that SCM + k-NN produced more subgroups than any other method, especially when compared to GMM and k-Means. This would seem to suggest that the method is more sensitive to subtler relationships between different combinations of socio-cultural and behavioral attributes.

As with the previous data set, we will also look at how similar subgroup assignments are between methods using the ARI scores, which are depicted in Fig. 2. Additionally, since a parliamentarian's stated faction is often used to define the factions within the parliament, we have also included the stated formal faction assignments as a subgrouping procedure for comparison. The ARI plot clearly demonstrates that GMM and k-Means produce the exact same subgroup assignments on this data set. This is good evidence that there is likely a good

Table 2 Summary statistics of subgroups found in Ukrainian Parliament data set

Subgrouping method	Number of subgroups	Avg. subgroup size	Std. subgroup size	Range of subgroup sizes	Silhouette
SCM+k-NN	40	13.05	8.09	[3,28]	0.18
Gaussian Mixture Model with BIC	2	261	124.45	[173,349]	0.37
k-Means with BIC	2	261	124.45	[173,349]	0.37
Affinity Propagation	29	18.0	10.43	[7,55]	0.10

split between two major subgroups within the parliament, but beyond those two major splits, finding subgroups becomes increasingly difficult. Also, no subgrouping compared well with the formal parties that the parliamentarians affiliate with. Thus, it would seem that while party is an important component in the subgroups of a political body like the Ukrainian Parliament, it is certainly not the only socio-cultural or behavioral component that influences actual subgroup formation. Lastly, while no subgrouping method is very similar in subgroup assignments, outside of GMM and k-Means, affinity propagation is the most similar to the other methods, on average. That would suggest that there is almost a hierarchy of subgrouping methods on this data set when it comes to the granularity of subgroups found. The methods of GMM and k-Means find very course, large subgroups and the methods of SCM + k-NN find very fine and nuanced subgroups. Affinity propagation find subgroups of granularity somewhere in between the other four methods.

As with previous data set, using SCM + k-NN also produces a latent graph which is visualized in Fig. 4. Most of the components in this network tend to form around

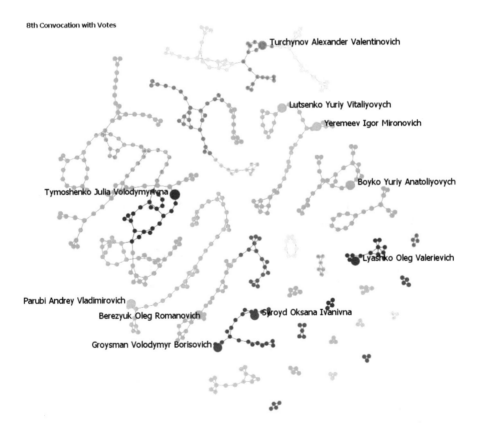

Fig. 4 Latent network of the Ukrainian Parliament data set. Coloring represents subgroup assignments, and key personalities are labeled in the network [22]

voting habits. The component containing Boyko Yuriy Anatoliyovych is routinely absent for or against most presidential bills, which would indicate that particular component is an opposition faction. And, while Boyko Yuriy Anatoliyovych is the leader of the "Opposition Bloc" political party, the members of his component are not all members of the "Opposition Bloc." That particular component also contains members of the "Revival" party and unaffiliated members of the parliament. Additionally, the component containing Turchynov Alexander Valentinovich, who is a member of the "People's Front" also did not participate in many of the presidential proposed votes. This component which is very inhomogenous in terms of party affiliation may represent an unofficial opposition.

Analyzing the network with standard social network analysis measures we observe some interesting results from the latent network. Dzhemilev Mustafa, Lopushansky Andriy Yaroslavovich, Bandurov Volodymyr Volodymyrovych, and Gerega Alexander Vladimirovich all score highly in betweenness centrality. Two of the four members are members of the Peter Porchenko Bloc and a different two are on the Committee on Energy Complex, Nuclear Policy and Nuclear Safety. All four of them seem to have varied voting records between them. As such, these individuals may represent important moderate elements of the parliament as they do not particularly ally with any faction but have behavior and attributes characteristic of many members of the parliament. Alekseev Sergey Olegovich, Artyushenko Igor Andreevich, Bublyk Yuri Vasilievich, and Boyko Olena Petrovna all score highly in degree centrality. Three of the four all come from the "Peter Porchenko Bloc." This would seem to indicate that the "Peter Porchenko Bloc" may wield a good deal of influence in the parliament due to the socio-central members of its party.

5 Discussion

The SCM + k-NN method finds a best-fit latent space for entities based on their socio-cultural attributes and behaviors and then finds a latent network and subgroup assignments by iteratively building k-NN graphs and evaluating the modularity of the subgroups and graphs. This method tended to produce smaller, more finely grained subgroups than any of the other methods we compared it to in this paper. In particular, for the Ukrainian Parliament data set, the SCM + k-NN found many more, and more nuanced subgroups than either GMM or k-Means when k was determined by BIC. This would suggest that the SCM + k-NN method would complement social science research in understanding populations from socio-cultural and behavioral data by finding nuanced subgroups and relationships between those subgroups. However, it also suggests that the method may be sensitive to noise present in the data, which is certainly a limitation in practical research. The sensitivity to noise in the data likely also affected the relatively high number of components; 8 for Hatfield–McCoy and 26 for the Ukrainian Parliament. If the method were less sensitive, there would be fewer components and the latent network would likely better illustrate community dynamics. Additionally, no distinction was made between socio-cultural attributes and behavioral attributes when performing SCM. This is a current limitation of the method and a direction for future research.

Our method also provides a novel method of constructing latent space networks. As opposed to traditional latent space models that are fit to data using maximum likelihood estimators, our proposed method fits a network to the data based on best fit distances and network characteristics (i.e., modularity). The links produced by this method tend to reflect neighborhoods of similarity of socio-cultural variables. Since the links only form to those entities who are nearby in the latent socio-cultural space, and only form when it supports local neighborhood development, measure by modularity, the latent network will emphasize neighborhoods of socio-cultural and behavioral similarity. Since socio-cultural similarity can be an important driver of link formation in social networks [9], these networks may then better represent a possible unobserved social network for a given population.

6 Conclusion and Future Work

Overall, our proposed method of SCM + k-NN produces meaningful subgroups on the two real-world data sets we analyzed. Usage on two real-world data sets demonstrated the utility of the subgroups, and latent network, that the method found. SCM + k-NN's performance in clustering was also different than existing unsupervised machine learning techniques. In particular, SCM + k-NN matched the performance of affinity propagation far more than it matched GMMs or k-Means. The end result of our proposed procedure is the possible relations and sub-communities that could exist in a population based upon just their shared behavior and socio-cultural attributes, which is useful for analysis in social science research.

One particular avenue of future research that we did not address in this paper is to use other unsupervised techniques with SCM. While we did show the utility of the k-NN network modularity maximization procedure to operate on SCMs with the SCM + k-NN method, it remains to be tried whether constructing the graph and subgrouping the graph based on other measure than modularity would work better with SCMs to create more connected latent graphs. Additionally, when using the SCM procedure there was no distinction between behavioral and socio-cultural variables; all variable contributed equally to the frequency counts. It would be interesting to adapt the SCM procedure in the future to somehow differentiate between these classes of variables. Finally, the data sets we considered for the case studies in this paper had a relatively modest number of attributes; 9 for the Hatfield–McCoy and 62 for the Ukrainian Parliament. It would be interesting to investigate how well the SCM procedure performs when we are able to obtain hundreds or thousands of attributes for each individual.

Acknowledgements This material is based upon work supported by the National Science Foundation Graduate Research Fellowship (DGE 1745016), Department of Defense Minerva Initiative

(N00014-15-1-2797), and Office of Naval Research Multidisciplinary University Research Initiative (N00014-17-1-2675). Any opinion, findings, and conclusions or recommendations expressed in this material are those of the authors and do not necessarily reflect the views of the National Science Foundation, Department of Defense, or the Office of Naval Research.

References

1. Levine, J.H., Carley, K.M.: SCM system. Technical Report. CMU-ISR-16-108, Institute for Software Research, Carnegie Mellon University (June 2016). http://reports-archive.adm.cs. cmu.edu/anon/isr2016/CMU-ISR-16-108.pdf
2. Morgan, G.P., Levine, J., Carley, K.M.: Socio-cultural cognitive mapping. In: Lee, D., Lin, Y.R., Osgood, N., Thomson, R. (eds.) Social, Cultural, and Behavioral Modeling, pp. 71–76. Springer International Publishing, Cham (2017)
3. Ruan, J.: A fully automated method for discovering community structures in high dimensional data. In: 2009 Ninth IEEE International Conference on Data Mining, pp. 968–973 (Dec 2009). https://doi.org/10.1109/ICDM.2009.141
4. Barabsi, A.L.: Network Science. Cambridge University Press, Cambridge (2015). http:// networksciencebook.com/
5. Snijders, T.A.B., Lazega, E.: Multilevel Network Analysis for the Social Sciences. Springer, Cham (2016)
6. Kim, B., Lee, K., Xue, L., Niu, X.: A review of dynamic network models with latent variables (2018). arXiv. https://arxiv.org/pdf/1711.10421.pdf
7. Hoff, P.D., Raftery, A.E., Handcock, M.S.: Latent space approaches to social network analysis. J. Am. Stat. Assoc. **97**(460), 1090–1098 (2002). https://doi.org/10.1198/016214502388618906
8. Leskovec, J., Lang, K.J., Mahoney, M.W.: Empirical comparison of algorithms for network community detection. CoRR abs/1004.3539 (2010). http://arxiv.org/abs/1004.3539
9. Newman, M.E.J.: Networks: An Introduction. Oxford University Press, Oxford (2010)
10. Schaub, M.T., Delvenne, J., Rosvall, M., Lambiotte, R.: The many facets of community detection in complex networks. CoRR abs/1611.07769 (2016). http://arxiv.org/abs/1611.07769
11. Jolliffe, I.: Principal Component Analysis. American Cancer Society (2005). https://doi.org/ 10.1002/0470013192.bsa501, https://onlinelibrary.wiley.com/doi/abs/10.1002/0470013192. bsa501
12. van der Maaten, L., Hinton, G.: Visualizing high-dimensional data using t-SNE. J. Mach. Learn. Res. **9**, 2579–2605 (2008)
13. Maier, M., Hein, M., von Luxburg, U.: Optimal construction of k-nearest neighbor graphs for identifying noisy clusters. Theor. Comput. Sci. **410**(19), 1749–1764 (2009). https://arxiv.org/ abs/0912.3408
14. Blondel, V.D., Guillaume, J.L., Lambiotte, R., Lefebvre, E.: Fast unfolding of communities in large networks. J. Stat. Mech. Theory Exp. **10** (2008). http://arxiv.org/abs/0803.0476
15. Pelleg, D., Moore, A.: X-means: extending k-means with efficient estimation of the number of clusters. In: In Proceedings of the 17th International Conference on Machine Learning, pp. 727–734. Morgan Kaufmann, San Francisco (2000)
16. Marin, J.M., Mengersen, K.L., Robert, C.: Bayesian modelling and inference on mixtures of distributions. In: Dey, D., Rao, C. (eds.) Handbook of Statistics, vol. 25. Elsevier, Amsterdam (2005). https://eprints.qut.edu.au/901/
17. Frey, B.J., Dueck, D.: Clustering by passing messages between data points. Science **315**(5814), 972–976 (2007). https://doi.org/10.1126/science.1136800, http://science. sciencemag.org/content/315/5814/972
18. Center for Computational Analysis of Social and Organizational Systems: ORA (June 2018). http://www.casos.cs.cmu.edu/projects/ora/

19. Hubert, L., Arabie, P.: Comparing partitions. J. Classif. **2**(1), 193–218 (1985). https://doi.org/10.1007/BF01908075,
20. Rousseeuw, P.J.: Silhouettes: a graphical aid to the interpretation and validation of cluster analysis. J. Comput. Appl. Math. **20**, 53–65 (1987). https://doi.org/10.1016/0377-0427(87)90125-7, http://www.sciencedirect.com/science/article/pii/0377042787901257
21. Center for Computational Analysis of Social and Organizational Systems: Public datasets (April 2018). https://www.casos.cs.cmu.edu/tools/datasets/external/index.php
22. Verkhovna Rada of Ukraine: 8th convocation (February 2018). http://rada.gov.ua/en

Bootstrapping the Gini Index of the Network Degree: An Application for Italian Corporate Governance

Carlo Drago and Roberto Ricciuti

Abstract We propose a new approach based on bootstrapping to compare complex networks. This is an important task when we wish to compare the effect of a (policy) shock on the structure of a network. The bootstrap test compares two values of the Gini index, and the test is performed on the difference between them. The application is based on the interlocking directorship network. At the director level, Italian corporate governance is characterized by the widespread occurrence of interlocking directorates. Article 36 of Law 214/2011 prohibited interlocking directorates in the financial sector. We compare the interlocking directorship networks in 2009 (before the reform) with 2012 (after the reform) and find evidence of an asymmetric effect of the reform on the network centrality of the different companies but no significant effects on Gini indices.

1 Introduction

During the past decades, many scholars have come up with theories to explain the presence of interlocking directorates (board members that simultaneously sit on more than one board: for a review of the literature see [1, 2]). From an economic standpoint, interlocking directorates are important because they can increase collusion among different companies whose directors sit on their respective boards, reducing consumer welfare. The effectiveness of "busy" board members sitting on several boards may also diminish, with less ability to check the chief executive officer's decisions, exposing companies to high risks [3]. Network analysis has been applied several times to the analysis of Italian corporate governance and ownership [4–9].

C. Drago
Niccolò Cusano University, Rome, Italy
e-mail: carlo.drago@unicusano.it

R. Ricciuti (✉)
University of Verona, Verona, Italy
e-mail: roberto.ricciuti@univr.it

© Springer Nature Switzerland AG 2020
G. Ragozini, M. P. Vitale (eds.), *Challenges in Social Network Research*,
Lecture Notes in Social Networks, https://doi.org/10.1007/978-3-030-31463-7_4

The chapter is organized as follows. In Sect. 2 we introduce our approach with simulations performed in Sect. 3. Section 4 presents our application, analyzing the effect of a reform banning interlocking directorates in banks and insurance companies in Italy, to see whether this led to significant changes in the structure of the network after the reform. Section 5 concludes.

2   The Approach

Given a network $G = (V, E)$ where V indicates the vertices of the network and E the edges, for each node we calculate the degree [10]:

$$C_D(v) = \deg(v) \tag{1}$$

The degree is the number of nodes that are their neighbors. In this way, each node is characterized by its local measure of centrality [11]. Then for each network, we compute the Gini inequality index [12]:

$$G_k = \frac{1}{n-1}\left(n+1-2\left(\frac{\sum_{j=1}^{n}(n+1-i)\,x_j}{\sum_{j=1}^{n} y_j}\right)\right) \tag{2}$$

where n is the number of nodes, x is the degree that characterizes the jth node. We have perfect equality if the Gini index is equal to 0, which means that all the degrees show the same value. For each network k we obtain different values for the Gini index.

The bootstrap test compares two Gini indices, G_1 and G_2, and tests the statistical significance of the difference $D = G_1 - G_2$ [13, 14]. It is important to note that we can bootstrap the distribution of G_1 or G_2 similarly to bootstrapping the distribution of D [13]. Following [13], it is important to note that the bootstrap distribution $\hat{F}(D)$ allows us to obtain the values for the hypothesis testing on D. With the bootstrap hypothesis testing method [15], we draw samples from the original data with replacement from data to obtain the bootstrap sample. Then we approximate the distribution of D by the bootstrap distribution of D_*. Finally, from the sampling distribution of D by obtaining their bootstrap estimate hypothesis testing [16, 17] can be carried out. Finally, the results are used to test whether the change in inequality in degree can be considered statistically significant. The code is written in Stata [18] and in R [19, 20].

3   Simulations

To try out the methodology, we simulated some networks on which we performed the algorithm to observe the changes that occur when impacted by structural changes.[1] We consider different networks and scenarios to test whether significant changes over time can be identified. The simulation is performed by randomly generating six networks and then executing the bootstrap test. In each test, we use two network topologies with the same number of nodes. In the first set, we start by considering two extreme cases: the first with a highly centralized network structure (a star) versus a structure evolved in an Erdös Renyi model [21] and the second a typically centralized (the Barabasi Albert model [22]) and a non-centralized network (the Erdös Renyi model).

We then provide further experiments by increasing the number of nodes/edges to construct more complex structures that may react differently to a shock. In this way, a more challenging environment is created for the null hypothesis. The second batch of simulations is based on the evolution of a network over time in which the deletion of some edges is simulated, followed by the addition of other edges in the second round. More precisely, we start from a Barabasi Albert Model and then randomly delete 4 edges in the first round and add 12 edges in the second. The initial network for both simulations is represented in Fig. 1. In the second experiment, we consider a higher additional number of edges by randomly removing and adding edges of the first network evolution simulation, i.e., 4 edges are deleted and 20 added.

We consider all possible connections on the nodes. The adjacency matrix is related to all possible connections which can occur in the network. Addition can occur randomly at each theoretically plausible connection and if the edge already exists, it is kept, otherwise, a new link is added. Therefore, the edge is added only where the connection is non-existent.

The final networks are shown in Fig. 2 (simulation 1) and Fig. 3 for simulation 2. It is interesting to note that very different network structures can be quickly obtained with the addition or deletion of the single edges. We simulate the network and compute the degree for each node, then calculate the Gini index for both networks.

Finally, we consider the third batch of simulations. We simulate the destruction (first case) and creation (second case) of a random number of edges between 100 and 1. We plot the initial networks (Figs. 4 and 5) and their final state after the destruction (Fig. 6) and the construction (Fig. 7) of new edges.

---

[1]In the simulations we have considered very general networks in which there are no effects caused by preferential attachment, triadic closure, clustering tendencies, constraints on the degree distribution due to transaction costs, which may characterize economic networks. The different additions/destructions of the links occur randomly. In the future, we will consider more complex cases that consider these features. These more complex structures may lead to complex reactions to the addition/destruction of the different nodes. In particular, the process of addition/destruction may be quicker or slower with the implication that the test may be more or less likely to reject the null hypothesis.

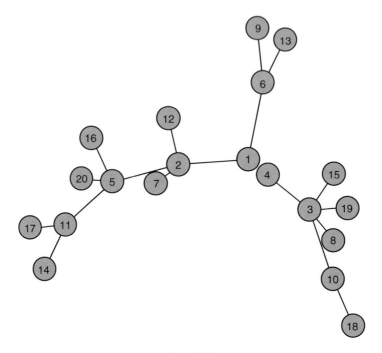

Fig. 1 Evolutive network simulations 1 and 2, initial state

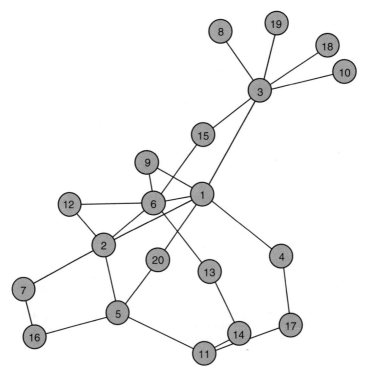

Fig. 2 Evolutive network 1, end of the simulation

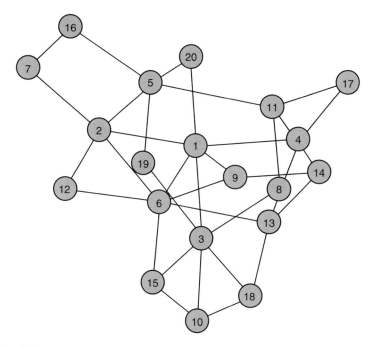

Fig. 3 Evolutive network 2, end of the simulation

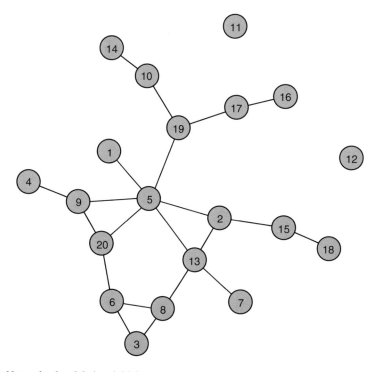

Fig. 4 Network edge deletion, initial state

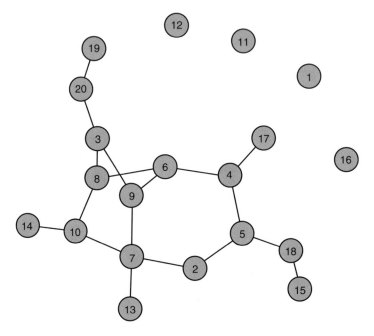

Fig. 5 Network edge creation, initial state

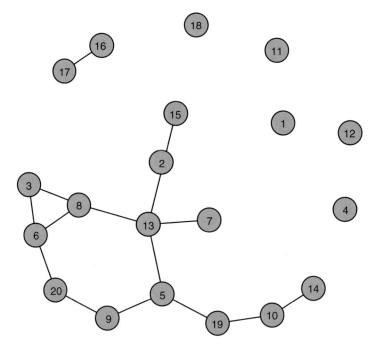

Fig. 6 Network edge deletion, end of the simulation

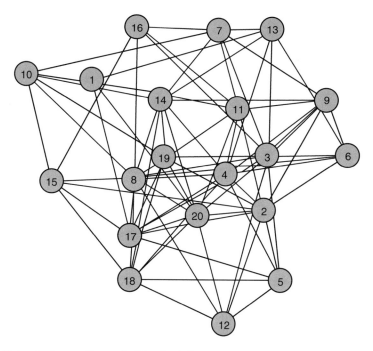

Fig. 7 Network edge creation, end of the simulation

Table 1 Simulation results

Coef.	Std. err.	Z	$P > |z|$
Centralized structure (star model) versus Erdös-Renyi model			
−0.19101	0.04908	−3.89	0.000
Model with similar structures (Barabasi-Albert models)			
−0.02894	0.08562	−0.34	0.735
Network evolution (1)			
−0.06268	0.05898	1.06	0.288
Network evolution (2)			
−0.11300	0.04575	−2.47	0.014
Network edge deletion			
0.07047	0.10186	0.69	0.489
Network edge creation			
−0.23972	0.07925	−3.02	0.002

Number of observations: 400, replications 2000

In our simulations (Table 1), the null hypothesis of a similar structure can be rejected in the first and simplest case but not in the second. In other words, in the first case we found a significant effect on the Gini index of the degree of the network, but not in the second case. In the two more complex simulations, the null hypothesis in Network evolution 1 cannot be rejected, whereas it may be slightly rejected in Network evolution 2. Finally, we are unable to reject the null hypothesis for network edge deletion, but we can reject the null for network edge creation.

4 Application

The application is based on the interlocking directorates network. Italian corporate governance features a high concentration of ownership and the presence of control-enhancing mechanisms that are conducive to controlling shareholders' dominance and exploitation of minorities. At the director level, corporate governance is characterized by the widespread recourse to interlocking directorates (directors sitting on more than one board at the same time, hereafter referred to as ID). Through cross-ownerships, circular ownerships, and interlocking directorates, the Italian system has been characterized by pyramidal groups headed by a small number of families that permanently control the firms.

A few reforms have been implemented over the last 15 years to open up the market for corporate control, reduce the scope for collusion, and to protect minorities from exploitation by controlling shareholders able to extract private benefits at the expense of the minority. The latest addition to this wave of reforms was a new measure introduced in 2011, article 36 banning interlocking, part of the Save Italy Decree which started life as Legal Decree 201/2011, published in Official Gazette on December 6, 2011. This decree was converted into law 214 with amendments in 2011 and published in the Official Gazette on December 27, 2011. Under article 36 paragraph 2b the requirement of 120 days to comply with the law ran from December 27, 2011. Therefore, the director of a bank or insurance company with incompatible appointments was required to choose one of the two (or more) positions by April 27, 2012, and failing this, would lose all the positions. The effects of the Law were in place when the data for our study were collected (December 31, 2012); hence, it is legitimate to compare 2012 with 2009 to see if the provision was effective in reducing ID in the financial sector. The reform aimed to break the ties between the sectors, increasing competition between financial companies. If this were true, we would observe a sparser network after the reform, with a significantly lower concentration in the Gini Index and more communities. Data were collected from listed companies in light of the Board of Directors for each firm on 31/12. Only the management board is considered for the few companies with the two-tier system. We used publicly available data from Consob (the Italian stock market regulator) and to collect the network data, we considered individual names and the related company and then created the two-way matrix, from which we were able to perform the one mode projection to obtain the adjacency matrices both for the network of directors and for the network of companies. A weighting represents the number of directors shared by connected companies.[2] Nodes represent companies. The isolates are companies that do not share any director with other companies.

---

[2]Working with the projected two-mode would make it possible to analyze the inequality in degree both on the company and the director side.

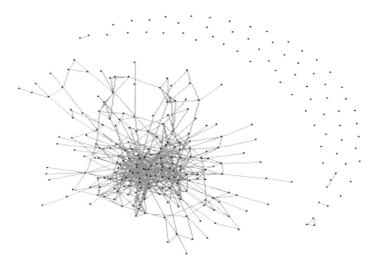

Fig. 8 The network in 2009

Fig. 9 The network in 2012

Figures 8 and 9 show the networks in the 2 years of interest. Visually comparing the structures of the networks, we find a group of nodes connected with each other on a first component and several isolates. Table 2 reports the descriptive statistics for the two networks.

We then perform a community detection analysis to decompose the networks in communities. We use the walktrap community detection approach [23] because it

Table 2 Descriptive statistics

	Network 2009	Network 2012
Nodes	278	251
Edges	576	387
Density	0.01496	0.012335
Islands\clusters	66	80
Global cluster coefficient	0.252179	0.260765
Diameter	10	15
Betweenness (mean)	225.4018	162.8853
Degree (mean)	4.143885	3.083665

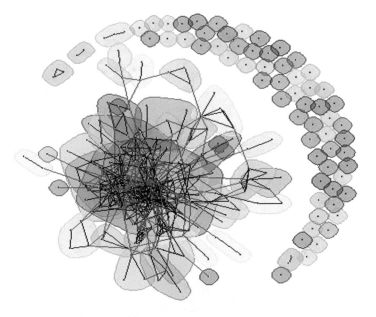

Fig. 10 Network in 2009: community structure

can detect and separate the different communities in this context. Walktrap is an appealing approach in our context because the random walks can be related to the dynamics of the information in the network.

The first community detection relating to the year 2009 (Fig. 10) shows 105 communities. Communities vary in size from 49 for the largest group to 1 for the isolates. The result shows that there are different groups of nodes strongly connected to each other. The communities tend to connect weakly compared to dense intra-community networks, and the network is utterly dissimilar to a random graph. Hence the expectation that the nodes have a different number of connections and different centrality in the same network. This result is particularly important because it confirms the need to consider the Gini index analysis to investigate the structure of the distribution of the degree over time.

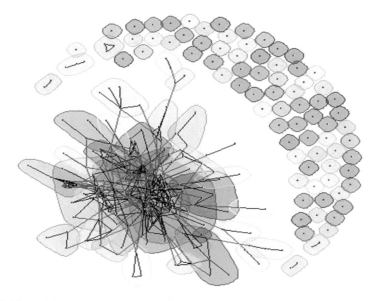

Fig. 11 Network in 2012: community structure

Regarding the community structure of the network in 2012 (Fig. 11), the number of communities increases to 111. There is greater fragmentation of the communities, with the largest community including 32 nodes against 1 for the isolates.

The heterogeneity in the distribution of the links between the nodes tends to decrease from 2009 to 2012 regarding the degree, for the first network it is in a range of 1 to 34, whereas in 2012 the range is from 1 to 21. In addition, the median and the mean are higher in 2009 (4.00 and 5.144, respectively) than in 2012 (3.00 and 4.084, respectively). Before the reform, the average value of the degree for each node is higher, possibly related to a different structure in the two networks. Finally, the variance of the degree for 2009 and 2012 moves from 21.71 to 13.43. Given the observed reduction in the heterogeneity of the degree, this may have an impact on its distribution, and this calls for an analysis of the Gini index of the degree in the 2 years.

The coefficient observed is 0.02 and the bootstrap standard error is 0.02575, with 2000 replications in the bootstrap process. Therefore, we detect no statistical significance, since we cannot reject the null hypothesis at 5%. In this case, we have included only the nodes which show a positive degree because we considered only nodes with at least a linkage, neglecting isolates. If we consider all the nodes, isolates, and non-isolates and repeat the analysis, we obtain a Gini index for the year 2009 of 0.55 and for 2012 of 0.60. These results are interesting because the deletion of some links due to the reforms has created an increase in the Gini index. The computed coefficient is 0.04151 with the bootstrap standard error equal to 0.02881. In this case too, we cannot reject the null hypothesis at the 5% significance level. We also plot the results obtained for the Lorenz Curves (Fig. 12) comparing the

 **Fig. 12** Lorenz Curves
(black line for 2009, dotted
line for 2012)

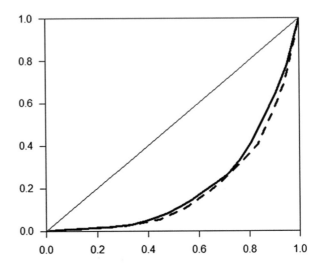

results for the year 2009 (black line) and for 2012 (dotted line). Overall, no strong
effect of the reform on the network can be found (as expected; see [6]), whereas an
asymmetric effect on the distribution of the edges (by considering both connected
nodes and isolated) is evident. The asymmetric effect is due to the presence of
some nodes that became more central in a local sense, because of edge deletion.
Interestingly, this potentially shows some unintended effects of the Law: by deleting
some edges, some nodes became even more important than before (see [24] showing
this effect in the period 1998–2006 for S&P MIB financial companies).

This result is similar to [25], where community detection techniques for the
analysis of the networks in 2009 and 2012 ascertained the effect of the reform on the
network of Italian directorates. They find that, although the number of interlocking
directorates decreases in 2012, the reduction takes place mainly at the periphery
of the network. The result is due to the fact the creation/deletion process fails to
activate the "structural change threshold."

5   Conclusions

This chapter presents a new method for the detection of statistically significant
changes in the network structure by a bootstrap test of the degree. The approach
appears to be very useful in analyzing the impact of exogenous shocks on a network
and in detecting impacts on the network structure. In our case, the shock was a
statutory amendment aimed at cutting the links between banking and insurance
companies in terms of directors sitting on more than one board. The methodology
can be extended to other node-level statistics, such as closeness, betweenness, and
so on. It is also possible to think about the centrality not only considering single
nodes but also groups of nodes (entire communities). The advantage of this approach

is the ability to observe the relative significance of shocks which can occur in a network system over time and space. The approach is also promising because it can be applied to other network structural measures such as betweenness. Possible problems can be identified in the fact that bootstrapping is not robust to outliers.

Our results show the limited effects of this legislative measure on the network of companies, possibly suggesting that a more far-reaching intervention was needed to achieve the desired outcome. Another possible reason could be that interlocking directorates are a symptom of cross-shareholding, and therefore regulation aimed at breaking these networks should address the former rather than the latter. Future research should try to understand the ultimate causes of the intertwining of listed Italian companies.

Acknowledgments We would like to thank Paolo Santella and Antonio Balzanella for useful discussions and two anonymous reviewers for constructive comments. Any remaining errors are ours alone. A previous version of the paper was presented at the SIS 2015 Conference.

References

1. Drago, C., Millo, F., Ricciuti, R., Santella, P.: Corporate governance reforms, interlocking directorship and company performance in Italy. Int. Rev. Law Econ. **41**, 38–49 (2015)
2. Enriques, L.: Corporate governance reforms in Italy: what has been done and what is left to do. Eur. Bus. Organ. Law Rev. **10**, 477–513 (2009)
3. Fich, E., Shivdasani, A.: Are busy boards effective monitors? J. Financ. **61**, 689–724 (2006)
4. Battiston, S.: Inner structure of capital control networks. Physica A. **338**(1–2), 107–112 (2004)
5. Battiston, S., Catanzaro, M.: Statistical properties of corporate board and director networks. Eur. Phys. J. B. **38**(2), 345–352 (2004)
6. Caldarelli, G., Catanzaro, M.: The corporate boards networks. Physica A. **338**(1–2), 98–106 (2004)
7. Bellenzier, L., Grassi, R.: Interlocking directorates in Italy: persistent links in network dynamics. J. Econ. Interac. Coord. **9**, 183–202 (2014)
8. Piccardi, C., Calatroni, L., Bertoni, F.: Communities in Italian corporate networks. Physica A. **389**(22), 5247–5258 (2010)
9. Vasta, M., Drago, C., Ricciuti, R., Rinaldi, A.: Reassessing the bank–industry relationship in Italy, 1913–1936: a counterfactual analysis. Cliometrica. **11**(2), 183–216 (2017)
10. Badham, J.M.: Commentary: measuring the shape of degree distributions. Netw. Sci. **1**, 213–225 (2012)
11. Wasserman, S.: Social Network Analysis: Methods and Applications, vol. 8. Cambridge University Press, New York (1994)
12. Forcina, A., Giorgi, G.M.: Early Gini's contributions to inequality measurement and statistical inference. Electron. J. Hist. Probab. Stat. **1**, 31 (2005)
13. Abdon   M.:   Bootstrapping   Gini.   http://statadaily.ikonomiya.com/2011/05/23/bootstrappinggini/
14. Mills, J., Zandvakili, S.: Statistical inference via bootstrapping for measures of inequality. J. Appl. Econ. **12**(2), 133–150 (1997)
15. Efron, B.: The Jackknife, the Bootstrap and Other Resampling Plans, vol. 38. Society for Industrial and Applied Mathematics, Philadelphia (1982)
16. Efron, B., Tibshirani, R.J.: An Introduction to the Bootstrap. CRC Press, Boca Raton (1994)

17. Biewen, M.: Bootstrap inference for inequality, mobility and poverty measurement. J. Econ. **108**, 317–342 (2002)
18. Jenkins, S.P.: INEQDECO: Stata module to calculate inequality indices with decomposition by subgroup. Statistical Software Components S366002, Boston College Department of Economics (1999), revised 22 Jan 2015
19. Csardi, G., Nepusz, T.: The igraph software package for complex network research. InterJournal Complex Syst. **1695**, 1 (2006). http://igraph.org
20. Iacus, S. M., Masarotto, G.: labstatR: Libreria del Laboratorio di Statistica con R. R package version 1.0.7. http://CRAN.R-project.org/package=labstatR (2012)
21. Erdös, P., Renyi, A.: On random graphs. Publ. Math. **6**, 290–297 (1959)
22. Barabasi, A.L., Albert, R.: Emergence of scaling in random networks. Science. **286**, 509–512 (1999)
23. Reiczigel, J., Zakari'as, I., R'ozsa, L.: A bootstrap test of stochastic equality of two populations. Am. Stat. **59**(2), 156–161 (2005)
24. Santella, P., Drago, C., Polo, A.: The Italian chamber of lords sits on listed company boards: an empirical analysis of Italian listed company boards from 1998 to 2006, MPRA Paper No. 2265 (2009)
25. Drago, C., Ricciuti, R.: Communities detection as a tool to assess a reform in the Italian interlocking directorship network. Physica A. **466**, 91–104 (2017)

Association Rules and Network Analysis for Exploring Comorbidity Patterns in Health Systems

Giuseppe Giordano, Mario De Santis, Sergio Pagano, Giancarlo Ragozini, Maria Prosperina Vitale, and Pierpaolo Cavallo

Abstract The presence of patients affected by different diseases at the same time is becoming a major health and societal issue. In clinical literature, this phenomenon is known as comorbidity, and it can be studied from the administrative databases of general practitioners' prescriptions based on diagnoses. In this contribution, we propose a two-step strategy for analyzing comorbidity patterns. In the first step, we investigate the prescription data with association rules extracted by a two-mode network (or bipartite graph) to find frequent itemsets that can be used to assist physicians in making diagnoses. In the second step, we derive a one-mode network of the diseases codes with association rules, and we perform the k-core partitioning algorithm to identify the most relevant and connected parts in the network corresponding to the most related pathologies.

Keywords Comorbidity pattern · Health data · Network data

1 Introduction

Today, the presence of patients affected by different diseases at the same time is becoming a major health and societal issue, because the presence of chronic or multi-morbid diseases is growing. In clinical literature, this phenomenon is known

G. Giordano (✉) · M. P. Vitale
Department of Political and Social Studies, University of Salerno, Fisciano (SA), Italy
e-mail: ggiordan@unisa.it

M. De Santis
Cooperativa Medi Service, Salerno, Italy

S. Pagano · P. Cavallo
Department of Physics E.R. Caianiello, University of Salerno, Fisciano (SA), Italy

G. Ragozini
Department of Political Science, University of Naples Federico II, Naples, Italy

© Springer Nature Switzerland AG 2020
G. Ragozini, M. P. Vitale (eds.), *Challenges in Social Network Research*,
Lecture Notes in Social Networks, https://doi.org/10.1007/978-3-030-31463-7_5

as comorbidity, that is, "any distinct additional entity that has existed or may occur during the clinical course of a patient who has the index disease under study" [12].

A possibility of studying comorbidity using available datasets resides mainly in administrative databases of general practitioners' (GPs') prescriptions. A recent literature review [28] showed that comorbidity can be studied from these databases based on either diagnoses, using the International Classification of Diseases (ICD) codes or pharmacy data. The comorbidity data are then described according to the co-occurrence of diagnostics in a single prescription as relationships between diagnoses.

In this scenario, the present contribution aimed at exploiting the GP administrative databases for health-care systems analysis, to mine the complexity of this information by using statistical techniques to extract the relevant features. A two-step strategy for analyzing comorbidity patterns is proposed based on association rules and social network analysis. In the first step, by considering the data structure as a transactional dataset, the prescription data can be investigated with the association rules extracted by the two-mode network to find frequent itemsets that can be used to assist physicians in making diagnoses. In the second step, the association rules can be represented as a one-mode network in which the diagnoses are the node, and the links are given by the rules. Thus, the one-mode network of the disease codes is analyzed by partitioning algorithms, to identify the most relevant and connected parts in the network structure corresponding to the most related pathologies.

This paper is organized as follows: Section 2 presents the theoretical framework. Section 3 introduces the data, and how they are used to define the comorbidity networks. Section 4 describes the strategy for the analysis. Section 5 reports the main findings, while Sect. 6 briefly discusses future lines of research.

2 Theoretical Background

Literature has differentiated two core concepts, comorbidity and multimorbidity. The latter is defined as the coexistence of two or more long-term conditions in an individual [19] not biologically or functionally linked, while the former concept, comorbidity, defines the coexistence of conditions that are actually linked [25], either biologically or functionally. If we define two different conditions, one called the "index condition" (IC) [5] and the other the "comorbid condition" (CC), in a state of comorbidity there will be one IC with one or more CCs. In contrast, as multimorbidity is usually defined as the co-occurrence of multiple chronic or acute diseases and medical conditions within one person without any reference to an index condition, in such a situation there will be a number of ICs and no CCs, but each IC could theoretically be the "seed" of a series of CCs.

Comorbidity, moreover, has been intended [20] in a positive or negative sense, namely syntropy and dystrophy [21]. The former is the mutual disposition, or the attraction of two or more diseases in the same individual, while the latter indicates those pathologies that are rarely found in the same patient at the same time. Thus, a given IC can attract one or more CCs to which it is syntropic, but at the same time, repulsing one or more different CCs to which the IC is dystropic. In the case of multimorbidity, two or more ICs not syntropic to each other can be syntropic or dystropic to a given number of CCs.

Time span and sequence [25] should be considered: a different length of co-occurrence and/or the appearance of the same two conditions in different sequences, e.g., *A* as the IC and *B* as the CC or vice versa, may imply different outcomes although the comorbidity conditions are the same. Simply considering this taxonomy, the intrinsically complex nature of comorbidity can be easily understood. If one considers the limitations of models and instruments used to represent this phenomenon, the following question becomes crucial [9]: "What do we observe when two disorders covary: a genuine phenomenon that is independent of our diagnostic criteria, measurement scales, and measurement models, or (in part) an artifact of the structure of these criteria and models?" In this sense, it has been suggested that research could usefully focus on behavioral medicine and secondary analyses of available datasets [24], and a standardized classification, such as the ICD, can be used as the basis to do that, taking into account the emerging construct of patient complexity.

On this basis, the morbidity and comorbidity burden is considered to be influenced by several factors, namely health-related characteristics and socio-economic, cultural, environmental, and behavioral characteristics, but there is a lack of agreement [5] on how to understand the complex interdependent relationships between diseases, due to the large number of variables (many of which are latent), the lack of accuracy in measurements, and the technological limitations in generating data.

For example, acute cardiovascular hospitalization Medicare claims data [8] have been successfully considered for understanding and measuring either the presence of comorbid conditions and of function-related indicators, such as depression, walking impairment. In addition, data from a network of GPs have been used to study disease clustering for a group of prevalent diseases, such as diabetes and osteoarthritis, showing a significant increment of the prevalence of comorbidity with respect to the patients' age.

In Italy, GPs can prescribe drugs, laboratory tests, imaging tests, specialist referrals, and hospitalization. For each type of prescription, a specific comorbidity network can be extracted by considering the co-occurrence of diagnostics in a single prescription as relationships between diagnoses. Thus, different patterns could emerge according to demographic or epidemiological factors. To the best of our knowledge, nobody has investigated these network aspects of comorbidity using an administrative GP database.

3   The Data

In the present contribution, the electronic health recordings (EHRs) of the prescriptions made by a group of 10 GPs are considered in the analysis, corresponding to a total number of 14,958 patients, and covering a time interval of 12 years, from 2002 to 2013. All the data used in the study were provided in an anonymous form, either for patients and GPs, according to the Italian law on privacy and guidelines of the Declaration of Helsinki. The study was retrospective observational, with anonymous data analyzed in aggregate form. The relevant ethics committee granted approval (Comitato Etico Campania Sud, document number 59, released on 2016-06-08). As the outcome of a patient's visit to a GP there is generally a prescription, containing a series of items of various types: drugs, laboratory tests, imaging tests, etc. The GP administrative prescription data used to provide, for each patient visit, the following information:

- patient ID, a unique random number assigned to the patient;
- demographic data, age and sex;
- prescription date;
- prescription type, drug, laboratory test, imaging, specialist referral, and hospitalization;
- prescription code, a specific code for each prescription type; and
- associated ICD diagnostic code, the pathology connected to the specific prescription.

The total number of analyzed prescriptions was 1,728,736, and their categorization by type is reported in Table 1.

Each item present in a GP prescription has, according to Italian National Health System rules, an associated possible disease encoded using the ICD, Ninth Revision, Clinical Modification (ICD-9-CM) [27]. The ICD is the standard diagnostic tool for epidemiology, health management, and clinical purposes, including the analysis of the general health situation of population groups. The simple count of codes is used to monitor the incidence and prevalence of diseases and other health problems, providing a picture of the general health situation of countries and populations. It has the general form $xxx.yy$, where xxx is the general disease, and yy is a specific occurrence, for example, 250 is the code for diabetes, and 250.91 is the code for

Table 1 Number of prescriptions by type

Prescription type	Number of prescriptions	Mean number of prescriptions per patient	Mean number of prescriptions per patient in 1 year
Drug	897,329	59.99	5.00
Lab	647,023	43.26	3.61
Others	184,384	12.32	1.02
Total	1,728,736	115.57	9.63

Table 2 ICD-9-CM groups

From	To	Acronym	Description
1	139	INFE	Infectious and parasitic diseases
140	239	NEOP	Neoplasms
240	279	META	Endocrine, nutritional and metabolic diseases, and immunity disorders
280	289	BLD	Diseases of the blood and blood forming organs
290	319	MENT	Mental disorders
320	359	NERV	Diseases of the nervous system
360	389	SENS	Diseases of the sense organs
390	459	CIRC	Diseases of the circulatory system
460	519	RESP	Diseases of the respiratory system
520	579	DIGE	Diseases of the digestive system
580	629	GEN	Diseases of the genitourinary system
630	679	PREG	Complications of pregnancy, childbirth, and the puerperium
680	709	SKIN	Diseases of the skin and subcutaneous tissue
710	739	MUSC	Diseases of the musculoskeletal system and connective tissue
740	759	CONG	Congenital anomalies
760	779	NEWB	Certain conditions originating in the perinatal period
780	799	ILL	Symptoms, signs, and ill-defined conditions
800	999	INJ	Injury and poisoning
E00	E99	EXT	External causes of injury
V00	V99	SUPP	Supplemental classification

diabetes type 1 (juvenile) with unspecified complication, not stated as uncontrolled. Only the general form, the first three digits, of ICD codes was used, grouped following the structure of the ICD-9-CM classification, to obtain 20 groups, each representing a specific epidemiological area. The 20 groups are listed in Table 2.

4 The Analytic Strategy

The comorbidity data are described according to the co-occurrence of diagnostics in a single prescription as relationships between diagnoses. At this end, a two-mode network is derived by considering the ICD-9-CM diagnostic codes and the prescriptions as two disjoint sets of nodes. The diagnoses are linked if corresponding codes appear in prescriptions for the same patient on the same day. The sex and age of patients, and the type and the time of the prescriptions, can be considered attributes of a given prescription. Specifically, in the present paper, we analyze men and women separately, as sex is a common risk factor in determining different pathology patterns. In this preliminary analysis, we consider only adult patients older than 35 years, without further differentiation.

More formally, the two-mode network can be represented as a bipartite graph $\mathcal{B}$ consisting of the two sets of relationally connected nodes, and can be represented by a triple $\mathcal{B}(\mathcal{D}, \mathcal{P}, \mathcal{A})$, with $\mathcal{D}$ denoting the set of ICD-9-CM codes, $\mathcal{P}$ the set of prescriptions, and $\mathcal{A} \subseteq \mathcal{D} \times \mathcal{P}$ the set of ties.

As we are interested in the comorbidity network, the two-mode network can be projected in a one-mode network *diagnoses* × *diagnoses* [11]. In such a case, two diagnoses are connected if they are present in the same prescription for the same patient, and the weight of the link is given by the number of patients in which this relationship is present. In Fig. 1 two different one-mode networks for women

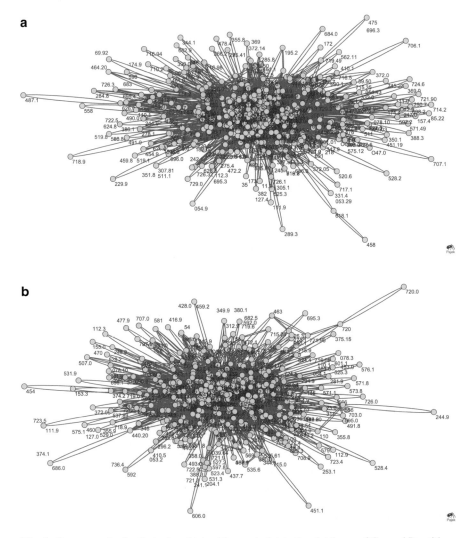

Fig. 1 One-network of pathologies obtained from administrative databases of General Practitioners prescriptions for women (**a**) and men (**b**)

and men, respectively, are plotted. The two networks are very dense, and present the well-known *hairball effect*. For this reason, it seems necessary to extract and explore the most interesting parts of the comorbidity network by considering a two-step strategy of analysis, as described below.

4.1 Association Rules Mining the Prescriptions Dataset

As the first step to disentangle the complexity in the comorbidity data, we extracted the association rules [2] from the two-mode network data. This technique is strictly related to the aim of finding frequent itemsets in a large dataset, and commonly applied in transactional data for marketing strategies.

Association rule mining from transaction data [2], *implication rules for market basket data* [4], and *recommendation technology* [1] are labels used in the literature to point out the search, visualization, and analysis of frequent itemsets in a large dataset. Transaction data arise in many business firms, in economic and commercial activity in daily operations. Association rules mining aims at discovering important relations between itemsets in the form of frequent items. For instance, an itemset can be the set of all products sold at a discount; on the other side, the set of buyers is the active set that operates by selecting items from the products set (the passive set). Such selections can be easily arranged in a table T where each row holds the elements of the active set (customers) and the columns are the passive set items (products). Each entry of this matrix indicates with 1 the selected product, and 0 otherwise. In this case, the interest could be in discovering whether subsets of products are bought together in a large percentage of cases, and derive suitable recommendations for similar buyers. This example drawn from a basket analysis situation can be generalized to different applicative fields.

Applying association rules in medical diagnosis can be used for assisting physicians to make diagnoses. Even if reliable diagnostic rules are difficult, and may result in hypotheses with unsatisfactory prediction, too unreliable for critical medical applications [14, 23] proposed a technique based on relational association rules and supervised learning methods. It helps to identify the probability of illness in a certain disease. In a study of protein structures, Gupta et al. [15] deciphered the nature of associations between different amino acids that are present in a protein. Such association rules are desirable for enhancing our understanding of the protein composition, and hold the potential to give clues regarding the global interactions among some sets of amino acids occurring in proteins.

Formally, let $\mathcal{D} = \{d_1, d_2, \ldots, d_i, \ldots, d_M\}$ be the set of all items (products or, in our case, either diagnoses or diseases); let $\mathcal{P} = \{p_1, p_2, \ldots, p_j, \ldots, p_N\}$ be the set of all transactions (prescriptions) where each transaction p_j is a subset of items chosen from $\mathcal{D}$, $(p_j \subset \mathcal{D})$, so that a prescription that had more diagnoses can be considered a transaction. A collection of items is termed an itemset. A prescription p_j contains an itemset X if X is a subset of p_j. The number of transactions that contain the particular itemset X defines the support count of X.

An association rule is an implication expression of the form $X \rightarrow Y$, where X and Y are disjoint itemsets of $\mathcal{D}$: $X \cap Y = \emptyset$, $X, Y \in \mathcal{D}$.

The strength of an association rule can be measured in terms of its *support*, *confidence*, and *lift*.

- The *support* of itemset $X \subset \mathcal{D}$ is the fraction of prescriptions in $\mathcal{P}$, of which X is a subset. It represents the prevalence of diagnoses (simple or multiple) in the universe of prescriptions $\mathcal{P}$, and can be stated as a relative frequency. A threshold s can be fixed so that when the *support* of a set of diagnoses is at least equal to the fixed value s, then the itemset is said to be frequent. The itemsets that exceed the minimum support s are said to be frequent patterns. These patterns can provide useful information about associations in prescriptions and diagnosis, in the case of simple itemset, that is, $(X \equiv d_i)$.
- The *confidence* of a rule C($d_1 \rightarrow d_2$) is the proportion of prescriptions with diagnosis d_1, in which diagnosis d_2 also appears. It is a conditional frequency, and can be stated in terms of the ratio between the *support* of the joint diagnoses $(d_1 \cup d_2)$ divided by the *support* of the condition diagnosis d_2.
- The *lift* of the association rule L($d_1 \rightarrow d_2$) is the ratio between the confidence of the rule and the product of the support of both sides of the rule d_1 and d_2. A *lift* greater than 1 means that diagnosis d_1 is likely to be present in a prescription if diagnosis d_2 is also present, while a *lift* less than 1 means that diagnosis d_1 is unlikely to be present in a prescription where diagnosis d_2 is present. A *lift* equal to 1 means that there is no association between diagnoses. These measures can be generalized to the case of multiple diagnoses; that is, the left side and/or the right side of the rule are itemsets of multiple diagnoses.

As the *confidence* does not take into account the prevalence of d_1, *confidence* C($d_1 \rightarrow d_2$) can be divided by the *support* of d_1.

Starting from this theoretical framework, our main purpose is to establish a link between association rules and network data, so that the transaction matrix T can be thought of as the affiliation matrix generated by a two-mode network or a bipartite graph $\mathcal{B}$, defined above. In the case of medical prescriptions, we consider a prescriptions per diagnosis affiliation matrix as bipartite graph $\mathcal{B} = (\mathcal{D}, \mathcal{P}, \mathcal{A})$, where $\mathcal{P}$ is the set of nodes representing the prescriptions; $\mathcal{D}$ is the set of nodes representing the diagnoses, and the edge $a_{i,j} \in \mathcal{A} \subseteq \mathcal{D} \times \mathcal{P}$ is established between p_j and d_i; that is, an edge exists if and only if prescription p_j contains diagnosis d_i. The collection of prescriptions per diagnoses, arranged in the binary affiliation matrix $\mathbf{A}$, is considered a transaction matrix where each prescription is a transaction defined on the set $\mathcal{P}$, so that $\mathcal{D}$ is the universal set of items, and each prescription in $\mathcal{P}$ is a subset of $\mathcal{D}$.

In this context, the aim stated in the association rules setting is to find sets of diagnoses that are strongly correlated in the prescription database, and are coherent with the notion of close neighbors in bipartite graphs; that is, pathologies directly connected to the same prescription. The common metrics for achieving this aim are the notions of *support*, *confidence*, and *lift*. These three measures induced by the association rules can be used to characterize the graph representation of the diagnosis itemset, defined as the projection of the bipartite graph induced by the prescription database. For the sake of computational efficiency, the analysis concentrates on the subset of diagnoses that have minimum *support s*.

However, if the graph is partitioned in k groups, it makes sense to apply the association rules search on the separate subgraph induced by the partition. The more meaningful rules are usually sorted by the decreasing value of *lift*. The first important rules among diagnoses will help uncover associations between frequent patterns of co-occurrence in diagnoses, in the whole set of prescriptions.

4.2 Network Analysis Tools

In the second step, being interested in the association of pathologies, we derived the one-mode network of the ICD-9-CM codes (diagnoses) by the association rules. The corresponding graph is represented by $\mathcal{G}(\mathcal{D}, \mathcal{E}, \mathcal{W})$, with $\mathcal{D}$ the set of ICD-9-CM codes, $\mathcal{E} \subseteq \mathcal{D} \times \mathcal{D}$ the set of edges, and $\mathcal{W}$ the set of weights. $w : \mathcal{E} \rightarrow \mathbb{R}$; $w(d_i, d_{i'})$ is the sum of the lift values of the rules with which the diagnoses are associated.

To identify the most relevant and connected parts of the network that correspond to the most related pathologies, we used the k-core partitioning algorithm [10, 22]. Given a graph $\mathcal{G}(\mathcal{D}, \mathcal{E}, \mathcal{W})$, a k-core is defined as a subgraph $\mathcal{H} = (\mathcal{C}, \mathcal{E}|\mathcal{C})$ induced by the subset $\mathcal{C} \subseteq \mathcal{D}$; $\mathcal{H}$ is a k-core if $\forall d \in \mathcal{C} : degree_{\mathcal{H}}(d) \geq k$. In addition, the subgraph $\mathcal{H}$ is the maximum subgraph with these characteristics.

The k-core procedure is then used to extract the relatively dense subnetworks (i.e., the maximum density subgraph of the k-core), that is, the subset of k-pathologies with the highest values of comorbidity occurrences, and to find cohesive subgroups of pathologies that are related by association rules with high *lift*. This network then is investigated with the usual exploratory tools of network analysis.

5 The Results

First, to mine association rules, we reduced the original database to specific prescriptions and patients' subpopulations as described below:

- Type = DRUGs which are commonly related to actual diagnosis;
- AGE range, from 35 to 110 years;
- SEX, *men* and *women*, with two separate databases; and
- removing non-relevant ICD-9-CM codes (Pregnancy, Congenital, Newborn, Ill-defined, etc.).

After the data were reduced, a set of 9845 patients (5252 women and 4593 men) were given 405,323 prescriptions (220,469 for women and 184,854 for men). The prescribed drugs totaled 627,924 (341,368 for males, and 286,556 for females). By applying the ICD-9-CM, we recognized a total number of 2387 diagnoses (1214 for women and 1713 for men), in a schematic view:

$$Patients \subset Prescriptions \subset Drugs \rightarrow Diagnoses : Patients \times Diagnoses.$$

Then, we set up a data matrix for men and one for women, in which the rows are patients, and the columns are diagnoses, to carry out separate analyses. Association rules were extracted according to the *Apriori* algorithm implemented in the *R* package *arules* [17] and visualized by *arulesViz* [16].

After a suitable setting of minimum support (0.01) and confidence (0.50), the algorithm extracted 517 rules for women and 175 rules for men plotted as scatterplots along their *support* and *confidence*, with the color intensity graded by *lift* in Fig. 2. Such graphical representation can be also interactively explored looking for more meaningful rules. An interactive data table visualization is given in Fig. 3 for the 10 most interesting rules. For the sake of simplicity, the results are decoded in Tables 3 and 4.

5.1   First Network Results

Given the 517 association rules for women and the 175 corresponding rules for men, we derived two different one-mode networks. We apply the k-core partitioning algorithm to extract the relatively dense subnetworks, and to find cohesive subgroups of pathologies that are related by the derived association rules with high *lift*.

Figure 4 portrays the two one-mode networks for women and men. The node is colored on the base of the k-core, while the node size is proportional to the betweenness centrality measures. Among the different network centrality measures [13], we selected the betweenness, because it helps in identifying the pathologies that are involved in many different rules, that is, in many comorbidity paths. In both figures, the red nodes correspond to the highest k-cores. It is possible to appreciate in both networks the diagnoses 401 and 462 are part of the central core, corresponding to *hypertension* and *acute pharyngitis*. Looking at the network for women, for example, the former is associated with some diseases that are directly connected, such as forms of *atherosclerosis* (414), *bronchitis* (491), *artery occlusion stenosis* (433), and *hypertensive heart disease* (402), and with some pathologies related to the aging process, such as *arthrosis* (715), and *cataract* (366), and to smoking, such as *chronic bronchitis* (491). In this network, an important node was given also by *cystitis* (595), which is also present in the network for men but with a lower betweenness centrality score. Looking at the network for men, for example, we notice that benign *prostatic hyperplasia* (600) appears in the highest k-cores.

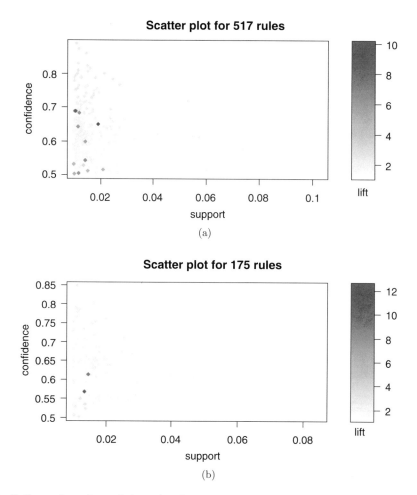

Fig. 2 Scatterplots of association rules along *support* and *confidence* with the color intensity graded by *lift* for women (**a**) and men (**b**)

6 Discussion and Concluding Remarks

Starting from the main findings, the first consideration involves the *lift* as an association rules' measure, and its interpretation in the study of comorbidity. As stated in Sect. 2, in the contribution of Puzyrev [21] about the comorbidity phenomenon, two aspects of comorbidity are presented: (1) *syntropy*, or direct comorbidity, which is defined as mutual disposition, or the attraction of two or more diseases in the same individual; (2) *dystrophy*, or inverse comorbidity, which indicates those pathologies that are rarely found in the same patient at the same time.

The *lift* appears to be a measure of these parameters, as a value greater than 1, meaning that diagnosis d_1 is likely to be present in a prescription if diagnosis d_2

LHS	RHS	support	confidence	lift ▾	count
All	All	All	All	All	All
[1] {477.0}	{464.1}	0.011	0.687	10.131	57.000
[24] {721.0}	{464.1}	0.019	0.650	9.585	102.000
[111] {466.0,464.1}	{493}	0.014	0.540	6.919	74.000
[83] {530.1,402.1}	{466.0}	0.010	0.688	6.711	55.000
[129] {530.1,493}	{466.0}	0.012	0.681	6.651	62.000
[130] {466.0,530.1}	{493}	0.012	0.504	6.457	62.000
[109] {530.1,464.1}	{466.0}	0.012	0.642	6.268	61.000
[110] {493,464.1}	{466.0}	0.014	0.597	5.826	74.000
[156] {464.10,535}	{715.90}	0.010	0.500	5.403	53.000
[82] {464.1,402.1}	{466.0}	0.010	0.530	5.174	53.000

Show 10 entries   Search:

Showing 1 to 10 of 517 entries   Previous 1 2 3 4 5 ... 52 Next

(a)

LHS	RHS	support	confidence	lift ▾	count
All	All	All	All	All	All
[4] {721.0}	{402.1}	0.014	0.568	12.593	63.000
[5] {721.0}	{464.1}	0.015	0.613	9.442	68.000
[65] {595,296.3}	{600}	0.012	0.549	3.245	56.000
[161] {715,595,401}	{600}	0.014	0.533	3.153	64.000
[123] {595,530.81}	{600}	0.014	0.520	3.076	64.000
[60] {600,414.8}	{595}	0.010	0.505	2.683	48.000
[170] {522.5,600,401}	{595}	0.012	0.500	2.655	55.000
[48] {250,715.90}	{401}	0.011	0.847	2.602	50.000
[107] {466,250}	{401}	0.017	0.800	2.456	76.000
[59] {250,414.8}	{401}	0.011	0.790	2.426	49.000

Show 10 entries   Search:

Showing 1 to 10 of 175 entries   Previous 1 2 3 4 5 ... 18 Next

(b)

Fig. 3 Interactive data table visualization for the ten most interesting rules for women (**a**) and men (**b**)

is also present, thus, measuring *syntropy*. However, a value lower than 1 measures *dystrophy*, because the co-presence of the left-hand diagnosis and the right-hand diagnosis appears less often together than expected. In other words, the lift values could be used as a quantitative measure of comorbidity and an indicator of its quality characteristic.

Another finding involves the sex effect. It clearly emerged that there are different association rules for men and women, as can be expected. Therefore, sex is a risk

Table 3 Association rules for diagnoses—women

	Left-hand rule → Right-hand rule
1	Rhinitis due to pollen → Acute tracheitis
2	Cervical spondylosis without myelopathy → Acute tracheitis
3	{Acute bronchitis; Acute tracheitis} → Asthma
4	{Esophagitis; Hypertensive heart diseas, benign} → Acute bronchitis
5	{Esophagitis; Asthma} → Acute bronchitis
6	{Acute bronchitis; Esophagitis} → Asthma
7	{Esophagitis; Acute tracheitis} → Acute bronchitis
8	{Asthma; Acute tracheitis} → Acute bronchitis
9	{Acute tracheitis without mention of obstruction; Gastritis and duodenitis} → Osteoarthrosis, unspecified whether generalized or localized
10	{Acute tracheitis; Hypertensive heart disease, benign} → Acute bronchitis

Table 4 Association rules for diagnoses—men

	Left-hand rule → Right-hand rule
1	Cervical spondylosis → Benign hypertensive heart disease;
2	Cervical spondylosis → Acute tracheitis
3	{Cystitis; Major depressive disorder, recurrent episode} → Hyperplasia of prostate
4	{Osteoarthrosis and allied disorders; Cystitis; Essential hypertension} → Hyperplasia of prostate
5	{Cystitis; Diseases of esophagus; Repair and plastic operations on joint structures} → Hyperplasia of prostate
6	{Hyperplasia of prostate; Other forms of chronic ischemic heart disease; Diseases of the Respiratory System} → Cystitis
7	{Periapical abscess; Hyperplasia of prostate; Essential hypertension } → Cystitis
8	{Diabetes mellitus; Osteoarthros NOS-unspec} → Essential hypertension
9	{Acute bronchitis and bronchiolitis; Diabetes mellitus} → Essential hypertension
10	{Diabetes mellitus; Chr ischemic hrt dis NEC} → Essential hypertension

factor that strongly affects association rules mining. The extraction of many more association rules for men than women, as evidenced by comparing Fig. 2a and b, and the different network structures in Fig. 4a and b, confirms previous findings [6, 7] on the greater complexity of the clinical framework of women. In this sense, the concept of "gender medicine" [3] should be taken into account for complex system studies. It is focused on the differences in pathophysiology, clinical aspects, prevention, and treatment of diseases in the genders, and it will have a deep impact on health policy, research, and teaching. Further developments will include the assessment of age-class effects as a risk factor, considering its interaction with sex, too.

Figure 4 has been reported in decoded form in Table 3 and Table 4, the entries are sorted by descending measure of Lift. The comparison of Tables 3 and 4 shows that there is a different complexity pattern of the ten most interesting rules between men and women, with the women (Table 3) showing a bigger presence of infectious disease related to respiratory and digestive system, and the men (Table 4) showing

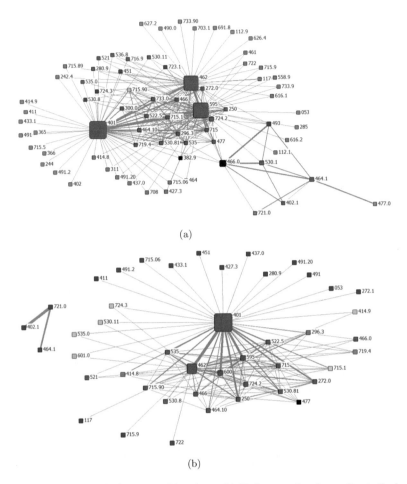

Fig. 4 One-mode networks for women (**a**) and men (**b**). Nodes are colored according to the k-core. Nodes' size is set proportional to the betweenness centrality measure

a more complex pattern of interactions, with a limited presence of infectious diseases. These findings suggest that there is also a qualitative gender difference in comorbidity, although further studies are strongly advised to confirm the finding and exclude sources of bias. This difference indicates that probably one interactome[1] is not enough, or at least, that we will need to understand how to navigate the

---

[1]The interactome [26] is the whole set of molecular interactions in a given cell, or individual, and it is represented as a graph of the biological network. It includes all the interactions between molecules belonging to different biochemical families, such as nucleic acids, proteins, lipids, carbohydrates, hormones, etc. Interactomics [18] is a discipline at the intersection of bio-informatics and biology that deals with the study of networks of interactions and their consequences.

interactome in different ways, according to the gender of the subjects under study. According to the findings of the present study, interactomics will probably need to add the dimension of gender differences to the field of study.

As future lines of research, the temporal dimension could be added to study the evolution of rules in the same cohort of patients over time. For instance, the *lift* metric could become an instrument of computational epidemiology, building software that automatically extracts data from a large database of EHRs. Such a monitoring instrument may measure the presence and variation of syntropy and dystrophy by gender and age, obtaining the trajectories of comorbidity in a given population. The trajectories, moreover, may be studied by plotting the value of *lift* for a given association in a cohort of patients followed during a certain number of years. The lift data could also be used for tracking population epidemiology relating to geographic, environmental, social, cultural, and other parameters, using a big data approach. A translational application could be a sort of computational epidemiology monitoring system. Finally, studies should also be performed using targeted data extraction approaches of EHRs, but this needs a much larger dataset. The more specific the analysis, the less numerous becomes the sample, and thus, the data extraction process strongly reduces the data yield.

References

1. Adomavicius G., Tuzhilin, A.: Toward the next generation of recommender systems: a survey of the state-of-the-art and possible extensions. IEEE Trans. Knowl. Data Eng. **17**(6), 734–749 (2005)
2. Agrawal, R., Imielinski, T., Swami, A.: Mining association rules between sets of items in large databases. In: Proceedings of the 1993 ACM SIGMOD International Conference on Management of Data, pp. 207–216. ACM, New York (1993)
3. Baggio, G., Corsini, A., Floreani, A., Giannini, S., Zagonel, V.: Gender medicine: a task for the third millennium. Clin. Chem. Lab. Med. **51**(4), 713–727 (2013)
4. Brin, S., Motwani, R., Ullman, J.D., Tsur, S.: Dynamic itemset counting and implication rules for market basket data. ACM SIGMOD Rec. **26**(2), 255–264 (1997)
5. Capobianco, E., Lio, P.: Comorbidity: a multidimensional approach. Trends Mol. Med. **19**, 515–521 (2013)
6. Cavallo, P., Pagano, S., Boccia, G., De Caro, F., De Santis, M., Capunzo, M.: Network analysis of drug prescriptions. Pharmacoepidemiol. Drug Saf. **22**, 130–137 (2013)
7. Cavallo P., Pagano, S., De Santis, M., Capobianco, E., General practitioners records are epidemiological predictors of comorbidities: an analytical cross-sectional 10-year retrospective study. J. Clin. Med. **7**(8), 184 (2018)
8. Chrischilles, E., Schneider, K., Wilwert, J., Lessman, G., O'Donnell, B., Gryzlak, B., Wright, K., Wallace, R.: Beyond comorbidity: expanding the definition and measurement of complexity among older adults using administrative claims data. Med. Care **52**(3), S75–S84 (2014)
9. Cramer, A.O.J., Waldorp, L.J., van der Maas, H.L.J., Borsboom, D.: Comorbidity: a network perspective. Behav. Brain Sci. **33**, 137–150 (2010)
10. De Nooy, W., Mrvar, A., Batagelj, V.: Exploratory social network analysis with Pajek, vol. 27. Cambridge University Press, Cambridge (2011)
11. Everett, M.G., Borgatti, S.P.: The dual-projection approach for two-mode networks. Soc. Netw. **35**, 204–210 (2013)

12. Feinstein, A.R.: The pre-therapeutic classification of co-morbidity in chronic disease. J. Chronic Dis. **27**, 455–468 (1970)
13. Freeman, L.C.: Centrality in social networks conceptual clarification. Soc. Netw. **1**(3), 215–239 (1978)
14. Gamberger, D., Lavrac, N., Jovanoski, V.: High confidence association rules for medical diagnosis. In: Proceedings of IDAMAP99, pp. 42–51 (1999)
15. Gupta, N., Mangal, N., Tiwari, K., Mitra, P.: Mining quantitative association rules in protein sequences. In: Williams, G.J., Simoff, S.J. (eds.) Data Mining, LNAI 3755, pp. 273–281. Springer, Berlin (2006)
16. Hahsler, M.: arulesViz: interactive visualization of association rules with R. R J. **9**(2), 163–175 (2017)
17. Hahsler, M., Chelluboina, S., Hornik, K., Buchta, C.: The arules R-package ecosystem: analyzing interesting patterns from large transaction datasets. J. Mach. Learn. Res. **12**, 1977–1981 (2011)
18. Kiemer, L., Cesareni, G.: Comparative interactomics: comparing apples and pears? Trends Biotechnol. **25**(10), 448–454 (2007)
19. Mercer, S.W., Smith, S.M., Wyke, S., O'dowd, T., Watt, G.C.: Multimorbidity in primary care: developing the research agenda. Fam. Pract. **26**, 79–80 (2009). Available via DIALOG. https://academic.oup.com/fampra/article/26/2/79/2367540. Cited 07 May 2018
20. Pfaundler, M., von Seht, L.: Uber Syntropie von Krankheitszustanden. Z. Kinderheilk. **30**, 298–313 (1921)
21. Puzyrev, V.P.: Genetic bases of human comorbidity. Genetika **51**, 491–502 (2015)
22. Seidman, S.B.: Network structure and minimum degree. Soc. Netw. **5**(3), 269–287 (1983)
23. Serban, G., Czibula, I.G., Campan, A.: A programming interface for medical diagnosis prediction. Stud. Univ. Babes-Bolyai Inform. **LI**, 21–30 (2006)
24. Valderas, J.M.: Increasing clinical, community, and patient-centered health research. J. Comorb. **3**, 41–44 (2013)
25. Valderas, J.M., Starfield, B., Sibbald, B., Salisbury, C., Roland, M.: Defining comorbidity: implications for understanding health and health services. Ann. Fam. Med. **7**, 357–363 (2009)
26. Vidal, M., Cusick, M.E., Barabási, A.L.: Interactome networks and human disease. Cell **144**(6), 986–998 (2011)
27. World Health Organization: International classification of Diseases (ICD) (2010). Available from: http://www.who.int/classifications/icd/en/
28. Yurkovich, M., Avina-Zubieta, J.A., Thomas, J., Gorenchtein, M., Lacaille, D.: A systematic review identifies valid co-morbidity indices derived from administrative health data. J. Clin. Epidemiol. **68**, 3–14 (2015)

A Mixture Model Approach
for Clustering Bipartite Networks

Isabella Gollini

Abstract This chapter investigates the latent structure of bipartite networks via a model-based clustering approach which is able to capture both latent groups of sending nodes and latent variability of the propensity of sending nodes to create links with receiving nodes within each group. This modelling approach is very flexible and can be estimated by using fast inferential approaches such as variational inference. We apply this model to the analysis of a terrorist network in order to identify the main latent groups of terrorists and their latent trait scores based on their attendance to some events.

1 Introduction

In recent years, there has been a growing interest in the analysis of network data. Network models have been successfully applied to many different research areas. We refer to [1] for a general overview of the statistical models and methods for networks.

In this chapter we will focus on finding clusters in a particular class of networks that is called bipartite networks. Bipartite networks consist of nodes belonging to two disjoint and independent sets, called sending and receiving nodes, such that every edge can only connect a sending node (e.g., actor) to a receiving node (e.g., event).

Latent variable models have been used to model the unobserved group structure of bipartite networks by setting sending nodes as observations and receiving nodes as observed variables (see, for example, [2, 3]). One important issue limiting the use of classical latent variable approaches, such as latent class analysis [4, 5] and stochastic blockmodels [6], is the assumption of local independence within the groups that, in the presence of a large heterogeneous network, may tend to yield

I. Gollini (✉)
University College Dublin, Belfield, Dublin, Ireland
e-mail: isabella.gollini@ucd.ie

© Springer Nature Switzerland AG 2020

G. Ragozini, M. P. Vitale (eds.), *Challenges in Social Network Research*,
Lecture Notes in Social Networks, https://doi.org/10.1007/978-3-030-31463-7_6

an overestimated number of groups making the results more difficult to interpret and potentially misleading. Aitkin et al. [4] proposed to use different models to overcome the issue of the local dependence assumption including the random Rasch latent class model in which they made use of class and event specific parameters that are, however, not able to capture the within class behaviour of each actor. Furthermore the computational effort required to estimate the model they propose is significant and this issue makes inference infeasible for large networks.

This chapter concerns the identification of groups in bipartite networks consisting of a set of actors and a set of events through a statistical mixture modelling approach which assumes the existence of a latent trait describing the dependence structure between events within actor groups and therefore capturing the heterogeneity of actors' behaviour within groups. This modelling framework allows for: model selection procedures for estimating the number of groups; explanation of the dependence structure of events in each group; description of the behaviour of each actor within each group by quantifying, through the latent trait, the conditional probability that a certain actor belonging to a certain group will attend a certain event. The posterior estimate of the latent trait scores can be visualized so as to interpret the estimated latent traits within each group. In order to fit the model variational inferential approaches are applied (see [7] and [8] for a comparison of estimates given by the variational and other approaches in latent trait models). The code implemented is included in the lvm4net package [9] for R [10]. The rest of this chapter is organized as follows: in Sect. 2 we describe the model and the inferential approach. In Sect. 3 we apply the proposed methodology to the Noordin Top terrorist bipartite network [4] in which we will aim to identify clusters of terrorists based on their attendance to a series of events in Indonesia from 2001 and 2010. We conclude in Sect. 4 with some final remarks.

2 Model-Based Clustering for Bipartite Networks

The relational structure of a bipartite network graph can be described by a random incidence matrix $\mathbf{Y}$ on N sending nodes (i.e., actors), R receiving nodes (i.e., events) and a set of edges $\{Y_{nr} : n = 1, \ldots, N; r = 1, \ldots, R\}$, where

$$Y_{nr} = \begin{cases} 1, & n \sim r; \\ 0, & n \not\sim r. \end{cases}$$

To cluster bipartite networks we adapt a flexible model-based clustering approach for categorical data, the mixture of latent trait analysers (MLTA) model introduced by Gollini and Murphy [8], to the context of bipartite network data. The MLTA model is a mixture model for binary data where observations are not necessarily conditionally independent given the group memberships. In fact, the observations within groups are modelled using a latent trait analysis model and thus dependence is accommodated. The MLTA model generalizes the latent class analysis and latent trait analysis by assuming that a set of N sending nodes can be partitioned into G groups, and the propensity of each actor to create links to the R receiving nodes depends on both the group they belong to and the presence of a D-dimensional continuous latent variable $\boldsymbol{\theta}_n$.

The model assumes that each sending node comes from one of G unobserved groups and defines $\mathbf{z}_n = (z_{n1}, z_{n2}, \ldots, z_{nG})$ as an indicator of the group membership, $z_{ng} = 1$ if actor n is from group g, with the following distribution:

$$\mathbf{z}_n \sim \text{Multinomial}(1, (\eta_1, \eta_2, \ldots, \eta_G)),$$

where η_g is the prior probability of a randomly chosen observation coming from group g ($\sum_{g'=1}^{G} \eta_{g'} = 1$ and $\eta_g \geq 0\ \forall g = 1, \ldots, G$). Further, the conditional distribution of $y_{n1}, \ldots, y_{nR}$ given that the observation is from group g is assumed to be a latent trait model with parameters b_{rg} and $\mathbf{w}_{rg}$.

Thus, the likelihood of the MLTA model is defined as

$$p(\mathbf{y}) = \prod_{n=1}^{N} \sum_{g=1}^{G} \eta_g\, p\left(y_{n1}, \ldots, y_{nR} | z_{ng} = 1\right)$$

$$= \prod_{n=1}^{N} \sum_{g=1}^{G} \eta_g \int p\left(y_{n1}, \ldots, y_{nR} | \boldsymbol{\theta}_n, z_{ng} = 1\right) p(\boldsymbol{\theta}_n)\, d\boldsymbol{\theta}_n,$$

where the conditional distribution of given $\boldsymbol{\theta}_n$ and $z_{ng} = 1$ is a Bernoulli distribution:

$$p\left(y_{n1}, \ldots, y_{nR} | \boldsymbol{\theta}_n, z_{ng} = 1\right) = \prod_{r=1}^{R} p\left(y_{nr} | \boldsymbol{\theta}_n, z_{ng} = 1\right)$$

$$= \prod_{r=1}^{R} \left(\pi_{rg}(\boldsymbol{\theta}_n)\right)^{y_{nr}} \left(1 - \pi_{rg}(\boldsymbol{\theta}_n)\right)^{1-y_{nr}},$$

and the response function for each group $\pi_{rg}(\boldsymbol{\theta}_n)$ is defined as the following logistic function:

$$\pi_{rg}(\boldsymbol{\theta}_n) = p\left(x_{nr} = 1 | \boldsymbol{\theta}_n, z_{ng} = 1\right)$$

$$= \frac{1}{1 + \exp\left[-(b_{rg} + \mathbf{w}_{rg}^T \boldsymbol{\theta}_n)\right]}, \quad 0 \leq \pi_{gr}(\boldsymbol{\theta}_n) \leq 1.$$

In addition, it is assumed that the D-dimensional latent variable $\boldsymbol{\theta}_n \sim \mathcal{N}(\mathbf{0}, \mathbf{I})$.

The attractiveness of receiving node r for sending nodes belonging to group g is modelled by the parameter b_{rg}. The parameter $\mathbf{w}_{rg}$ measures the heterogeneity of the behaviour of sending nodes belonging to group g to connect to the receiving node r (i.e., the heterogeneity of terrorists belonging to the latent group g in attending event r); it also accounts for the dependence between receiving nodes. The vector $\boldsymbol{\theta}_n$ contains the latent variables explaining the propensity of forming links for sending node n, i.e., the propensity of terrorist n to attend the events.

We also use a constrained model with common variable-specific slope parameters across groups (i.e., $\mathbf{w}_{rg} = \mathbf{w}_{rg'} = \mathbf{w}_r$, where $g \neq g'$):

$$\pi_{rg}(\boldsymbol{\theta}_n) = \frac{1}{1 + \exp\left[-(b_{rg} + \mathbf{w}_r^T \boldsymbol{\theta}_n)\right]}, \quad 0 \leq \pi_{gr}(\boldsymbol{\theta}_n) \leq 1,$$

This model is particularly useful to avoid the estimation of too many parameters, especially when the data set is complex, with actors coming from several latent groups and the continuous latent variable having high dimensionality.

The likelihood of the MLTA model is computationally intractable. For this reason [8] proposed to use a double EM algorithm with variational approximation of the likelihood to fit this model, also guaranteeing fast convergence. The main aim of this variational approach is to maximize the Jaakkola and Jordan [11] lower bound of the likelihood function. This lower bound is a function of auxiliary parameters, called variational parameters, that are optimized to tighten this lower bound. The standard errors of the model parameters can be calculated using the jackknife method [12]. For full details of the double EM algorithm, we refer to [8].

Since the EM approach is adopted, there is the issue that the results may be affected by the risk of converging to a local maximum instead of the global maximum approximate likelihood. For this reason, it is generally advisable to run the algorithm several times using different initializing values, and select the solution with maximum approximate likelihood. The application of the variational approach makes the estimation procedure much more efficient than most of the classical simulation-based estimation methods even when multiple starts are employed.

However, the approximation of the log-likelihood obtained by using the variational approach with the Jaakkola and Jordan lower bound is always less or equal than the true log-likelihood, so before performing model selection based on the likelihood, like the Bayesian Information Criterion (BIC) [13], it may be advantageous to get a more accurate estimate of the log-likelihood at the last step of the algorithm using Gauss–Hermite quadrature [8].

3 Noordin Top Terrorist Network

The Noordin top terrorist network data [14] displayed in Fig. 1 is a bipartite network oriented around the Malaysian Muslim extremist Noordin Mohammad Top (ID: 54) and his collaborators (the data set is available in the manet package [15] for R). The data include relational information on $N = 79$ sending nodes that are individuals belonging to terrorist/insurgent organizations and on $R = 45$ receiving nodes that

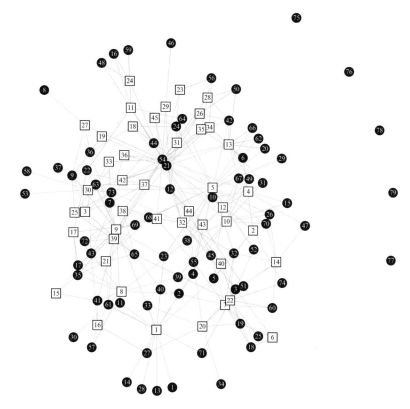

Fig. 1 Noordin bipartite graph. Black circles indicate the sending nodes (i.e., terrorists), the white squares indicate the receiving nodes (i.e., events)

represent events in Indonesia and nearby areas from 2001 to 2010. The incidence matrix contains links encoding the attendance behaviour of the terrorists to the events.

3.1 Statistical Analysis

We apply the MLTA modelling approach to the Noordin Top Terrorist Network. To avoid the issue of getting estimates affected by convergence to a local maximum, we use ten random starts of the algorithm and only the estimates corresponding to the maximum likelihood value are selected. The model parameters b_{rg} and $\mathbf{w}_{rg}$ are initialized by random generated numbers from a $\mathcal{N}(\mathbf{0}, \mathbf{I})$ and the variational parameters are initialized to be equal to 20 in order to reduce the dependence of the final estimates on the initializing values.

Table 1 BIC results for standard and constrained MLTA models with different number of groups and dimensions

	$D = 0$	$D = 1$		$D = 2$		$D = 3$	
		Common $\mathbf{w}_r$		Common $\mathbf{w}_r$		Common $\mathbf{w}_r$	
$G = 2$	2062	2138	**2034**	2389	2096	2793	2311
$G = 3$	2157	2403	2115	2876	2229	3417	2434
$G = 4$	2290	2730	2249	3385	2385	4419	2595

Value in bold denotes the smallest BIC value

The model is fitted on a range of groups, from 2 to 4 and the continuous latent variable takes value D from 0 to 3. For $D = 0$ the MLTA model reduces to a latent class analysis where the observations are assumed to be conditionally independent given the group membership. Model selection is performed on both the unconstrained MLTA and the constrained model with common slope.

The Bayesian Information Criterion (BIC) [13] is used to select the best model, and it is defined as:

$$\mathrm{BIC} = -2\ell_{\mathrm{GH}} + k\log(N),$$

where ℓ_{GH} is the estimate of the log-likelihood at the last step of the algorithm obtained by using Gauss–Hermite quadrature, k is the number of free parameters in the model and N is the number of sending nodes. The model with the lower value of BIC is preferable.

Table 1 shows the BIC values for models with increasing dimensionality. The best model selected is the one with two groups, a one-dimensional latent trait and common slope across groups.

For the best model selected, the values of the mixing proportions are: $\eta_1 = 0.57$ (SE = 0.080) for Group 1, and $\eta_2 = 0.43$ (SE = 0.084) for Group 2.

3.2   Interpreting the Actor's Behaviour

The sending nodes are partitioned into the two groups according to their maximum a posteriori (MAP) probability that they belong to each group. Figure 2 shows the posterior probability of each actor to belong to each group.

Most of the terrorists have been assigned to a particular group with probability very close to 1. In particular, Noordin Top (ID 54), attending 23 events, and Azhari Husin (ID 21), attending 17 events, are allocated together into Group 1 with probability 1. The 'lone wolves' (IDs 75, 76, 77, 78, 79), i.e., terrorists who have not attended any event, have been assigned to Group 1, but the uncertainty associated with their group membership is very large: in fact, their posterior probability to belong to Group 1 is 0.6.

In order to have a deeper understanding of group memberships we can use the information provided by the posterior distribution of the latent trait score θ_n conditional on the observation belonging to a particular group which can be obtained from the model estimates (see Fig. 3).

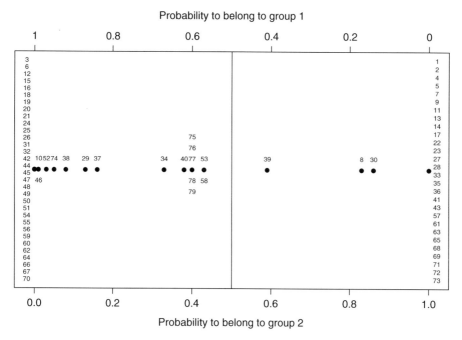

Fig. 2 Probability to belong to a group for each terrorist in the best model selected (two groups, a one-dimensional latent trait and common slope across groups)

The posterior mean estimates of these θ_n together with the information about event attendance y_{nr} can be used to interpret the latent variables within each group: Fig. 4 allows us to notice that in Group 1 the terrorists with high values went to events 7, 14 and 22 and most of the terrorists with low values went to events 13, 26, 34, 42. In Group 2 positive values are assigned to those terrorists who attended event 1 (it is also possible to notice that none of the terrorist in Group 1 attended event 1), negative values of the latent trait are associated with terrorists who attended events 2, 9, 25 and 33.

3.3 Interpreting the Events Attendance

A measure of the heterogeneity of attending event r within group g is given by the slope value $\mathbf{w}_{rg}$; the larger the value of $\mathbf{w}_{rg}$, the greater the differences in the probabilities of sending a link (going to event) r for actors from group g.

The choice of a model with the common slope ($w_{r1} = w_{r2} = w_r$) in all groups suggests the latent trait has the same effect in all groups. From Fig. 5 it is possible to notice that most of the slope parameters are non-zero, meaning that the latent trait introduces significant variation within the groups. This indicates that there is

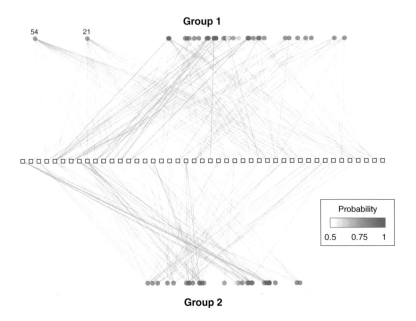

Fig. 3 Latent structure of the bipartite network. The terrorists are partitioned into the two groups according to their maximum a posteriori (MAP) probability and plotted according to their latent trait position. The darker the vertex colour, the higher their MAP of belonging to the group

considerable variability within the event attendance in the two groups and that some events are positively dependent (i.e., those going/not going to one event will tend to be going/not going in the others events) and others are negatively dependent. The dependence between events r and k in group g is given by $\mathbf{w}_{rg}^T\mathbf{w}_{kg}$, and the results are shown in Fig. 6. Red (blue) squares in the heatmap mean positive (negative) dependence, the darker they are, the higher is the dependence between two events. Figure 6 shows that the two set of events $(1, 6, 7)$ and $(3, 33, 36)$ are positively dependent within them and negatively between them.

The heatmap displayed in Fig. 7 represents the values of the log{*lift*} [16] that can be used to quantify within each group the effect of the dependence on the probability of attending two events compared to the probability of attending two events under an independence model. Mathematically the log{*lift*} for events r and k for actors belonging to group g is defined as

$$\log\{\textit{lift}\} = \log\left\{\frac{\mathbb{P}\left(y_{nr} = 1, y_{nk} = 1|z_{ng} = 1\right)}{\mathbb{P}\left(y_{nr} = 1|z_{ng} = 1\right)\mathbb{P}\left(y_{nk} = 1|z_{ng} = 1\right)}\right\},$$

where $r = 1, 2, \ldots, R$ and $r \neq k$. Two independent events have log{*lift*} $= 0$: the more two events are positively dependent, the higher the value of the log{*lift*}. Lift

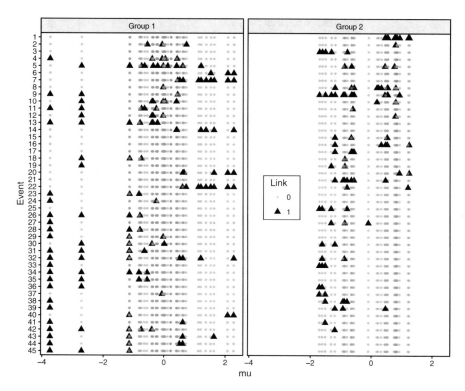

Fig. 4 Posterior mean estimate of the latent trait scores θ_n within each group for each actor and attendance to event

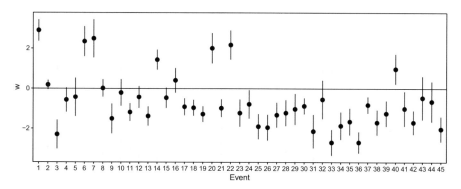

Fig. 5 Estimates of slope parameters for each receiving node (event) in the network and associated 95% confidence interval

values that are much less than 0 provide evidence of negative dependence within groups. Figure 7 shows that in Group 1 there is high negative dependence between events 1 and events 3, 33, 36, and in Group 2 there is high positive dependence between the events 27, 32, 34, 35, 45.

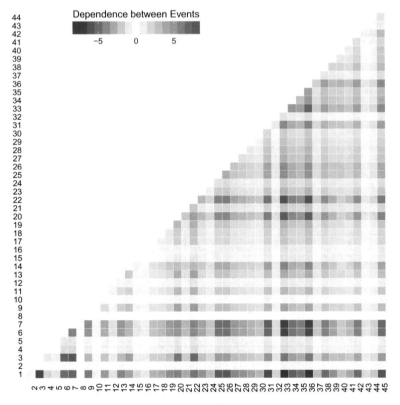

Fig. 6 Dependence between events, calculated as $\mathbf{w}_r^T \mathbf{w}_k$

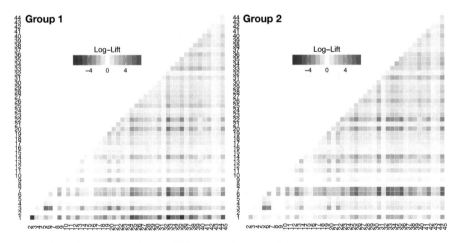

Fig. 7 Log-Lift for each pair of receiving nodes (events) of the network

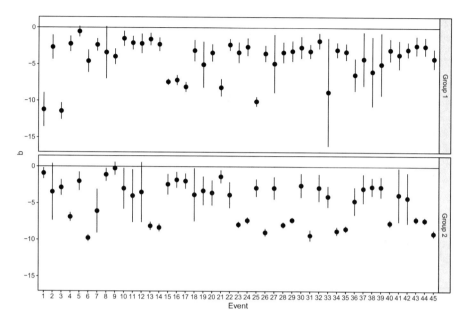

Fig. 8 Estimates of the intercept parameters for each receiving node (event) and corresponding 95% confidence intervals

The attractiveness of event r for actors belonging to group g is modelled by b_{rg}. Figure 8 shows that most of the values are significantly negative highlighting the sparse structure of the network.

Since $\theta_n \sim \mathcal{N}(0, 1)$, the probability that the median individual in group g attends events r can be calculated from the attractiveness parameters through the relationship:

$$\pi_{rg}(0) = p(x_{nr} = 1 | \theta_n = 0, z_{ng} = 1) = \frac{1}{1 + \exp(-b_{rg})}.$$

From Fig. 9 it is evident the different behaviour of the actors belonging to the two groups. Actors in Group 1 have high probability to attend events 7, 13, 14, 43, 44, 45, while those in Group 2 have very low probability to attend these events ($<10^{-04}$), and none of the terrorists in the data set assigned to Group 2 actually attended those events. Similarly Actors in Group 2 have high probability to attend events 1, 3, 15, 16, 17, 21, while those in Group 1 have very low probability to attend these events ($<10^{-04}$), and none of the terrorists in the data set assigned to Group 1 actually attended those events. Overall the probability that the median actor in each group attends any event is quite low (the highest probability of 0.438 for event 9 in Group 2). This is due to the fact that the number of terrorists attending the same events ranges from a minimum of 3 up to a maximum of 18 out of the total of 79 terrorists.

The proposed methodology is implemented in the **lvm4net** package for R.

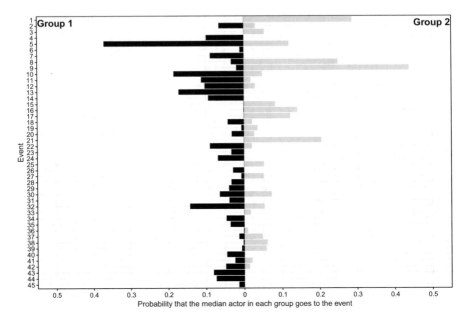

Fig. 9 Probability that the median actor in each group goes to the event

4 Conclusions

In this chapter, we have presented an application of a finite mixture model to the clustering of bipartite network data. The modelling framework is particularly flexible and useful for describing the between-group structure using a discrete latent variable and the within-group structure using a continuous latent variable. We have also illustrated how a variational inferential approach can be adopted to estimate the model efficiently. The model has been employed to analyse the relational connectivity patterns of the Noordin Top terrorist network. This has allowed us to find two main groups of terrorists based on their attendance to some events and yield important insights about both terrorists' behaviour within each group and the amount of dependence between events attended by them.

References

1. Salter-Townshend, M., White, A., Gollini, I., Murphy, T.B.: Review of statistical network analysis: models, algorithms, and software. Stat. Anal. Data Min. **5**(4), 243–264 (2012)
2. Aitkin, M., Vu, D., Francis, B.: Statistical modelling of the group structure of social networks. Soc. Netw. **38**, 74–87 (2014)
3. Ranciati, S., Vinciotti, V., Wit, E.C.: Identifying overlapping terrorist cells from the Noordin top actor-event network (2017). arXiv preprint:171010319

4. Aitkin, M., Vu, D., Francis, B.: Statistical modelling of a terrorist network. J. R. Stat. Soc. Ser. A Stat. Soc. **180**(3), 751–768 (2017)
5. Bartholomew, D.J., Knott, M., Moustaki, I.: Latent Variable Models and Factor Analysis: A Unified Approach, 3rd edn. Wiley, Hoboken (2011)
6. Nowicki, K., Snijders, T.A.B.: Estimation and prediction for stochastic blockstructures. J. Am. Stat. Assoc. **96**, 1077–1087 (2001)
7. Tipping, M.E.: Probabilistic visualisation of high-dimensional binary data. In: Proceedings of the 1998 Conference on Advances in Neural Information Processing Systems 11, MIT Press, Cambridge, pp. 592–598 (1999)
8. Gollini, I., Murphy, T.B.: Mixture of latent trait analyzers for model-based clustering of categorical data. Stat. Comput. **24**(4), 569–588 (2014)
9. Gollini, I.: lvm4net: Latent Variable Models for Networks (2019). https://CRAN.R-project. org/package=lvm4net
10. R Core Team: R: A Language and Environment for Statistical Computing. R Foundation for Statistical Computing, Vienna (2019). http://www.R-project.org
11. Jaakkola, T.S., Jordan, M.I.: A variational approach to Bayesian logistic regression models and their extensions. In: Proceedings of the Sixth International Workshop on Artificial Intelligence and Statistics (1996)
12. Efron, B.: Nonparametric estimates of standard error: The jackknife, the bootstrap and other methods. Biometrika **68**(3), 589–599 (1981)
13. Schwarz, G.: Estimating the dimension of a model. Ann. Stat. **6**(2), 461–464 (1978)
14. Everton, S.F.: Disrupting Dark Networks, vol. 34. Cambridge University Press, Cambridge (2012)
15. Ranciati, S., Vinciotti, V., Wit, E.: manet: Multiple Allocation Model for Actor-Event Networks (2018). https://CRAN.R-project.org/package=manet
16. Brin, S., Motwani, R., Ullman, J.D., Tsur, S.: Dynamic itemset counting and implication rules for market basket data. SIGMOD Rec. **26**(2), 255–264 (1997)

A DEA-Based Network Formation Model. Micro and Macro Analysis

Claudio Pinto

Abstract Empirical literature shows that networks occupy an important place in a variety of economic phenomena. The first segment of related research considers the network as a date and then studies its phenomena, while the second deals with how and why networks are formed. The current study falls into this second vein and proposes a network formation model based on DEA. This work exploits the DEA methodology as a process generating relational data. In other words, the relational variable, which is the empirical basis of a network, is generated within a DEA model. In this work, a strategic economic model is developed in which the usefulness of the agents depend exclusively on direct links and their formation depends unilaterally on the decision of a single agent within the dyad. From the statistical point of view, we will propose an independent dyad model in which each dyad has a dichotomous state. We will then go on to define equilibrium as couple stability for the DEA-based network. Finally, we will conduct a micro and macro analysis of the network, estimate an ERGM based on our proposed statistical model, and compare the properties of some standard statistical models with those of our DEA-based network based on simulations.

1 Introduction

In empirical literature, networks play an important role in many economic phenomena. The research on this subject unfolds in two strands. The first strand considers the network as a date and studies the involved economic processes (e.g., social learning, search for a job, effects of peers in the education, and so on). In comparison, the second strand deals with why and how networks (social and economic) are formed [18, 46] for example, it reviews the literature on economic consequences of the structure of social networks and develops a taxonomy of

C. Pinto (✉)
University of Salerno, Fisciano (SA), Italy
e-mail: clpinto@unisa.it

© Springer Nature Switzerland AG 2020   93
G. Ragozini, M. P. Vitale (eds.), *Challenges in Social Network Research*,
Lecture Notes in Social Networks, https://doi.org/10.1007/978-3-030-31463-7_7

micro and macro characteristics of social iterations. At the macro level, the authors indicates the distribution of degrees and the average grades as network features that have important consequences on individual behavior, thus, underlining the consequences on diffusion processes within the network. At the micro level for [46], there are three perspectives: (1) clustering and support, (2) centrality measures, and (3) multiple networks and strength of relationships. For example, the analysis of the centrality of a node in a network tells us about the potential power of the node, its influence or prestige, and so on. There are four major classes of centrality measures [48], and we will use some of them in our micro analysis. The macro analysis will consider the density of the network, the analysis of the triads, the network efficiency, and some other aspects. In this paper, we propose a DEA model for a DMU sample (see Appendix 1) and use the optimal lambda matrix to construct an adjacency matrix (see Appendix 2) to represent the DEA-based network. We will then conduct micro and macro analyses to understand its characteristics. In brief, the study involves the development of an economic model of network formation based on the DEA of a strategic nature, characterization of its equilibrium, development of an independent statistical dyad model, estimation of an ERGM model based on the specification of the statistical model, and some simulations. We believe that the paper is innovative in that it considers the DEA methodology as a generator of relational data and strategies for the strategic economic model. The paper is structured as follows. In Sect. 2, we present the works related to the topic. In Sect. 3, we develop the strategic model, propose the concept of couple stability, and clarify the network formation mechanism. In Sect. 4, we offer a micro and macro analysis of DEA-based network building in virtual DMU sample. In Sect. 5, we present the estimation of an ERGM model and the simulations with relative comparisons. Finally, in Sect. 6, we present our discussion and our conclusions.

2   Related Works

In empirical literature a first strand of research considers to establish the network and study the economic phenomena operating in it as happened for example with social learning, research in the labor market [1–3], peer effect in education [4], adoption of technology [5], and so on. Another strand of literature considers why and how economic networks are formed as for example in [6, 7]. In these latter templates (for a review see [8, 9]) agents create links according to specific rules, for example decisions can be taken simultaneously (i.e., [12, 22]), or sequentially (i.e., [11]), unilaterally (i.e., [7]), or bilaterally (i.e., [6]); the information may be complete (i.e., [12, 13]) or incomplete (i.e., [14, 15]); an explicit concept of equilibrium may be offered (i.e., Nash equilibrium, Bayesian Nash equilibrium, Nash stability [17], pairwise stability [6], pure strategy pairwise Nash equilibrium (among others [3, 6, 17]); take transferable or non-transferable payoffs according to the actions of the players and their characteristics. Conversely, econometric issues related to network formation are taken into account by [18] and relate to dyad-based models which may also consider externalities, i.e., [19] (in many economic contexts decisions to form links are interdependent and the presence of a link in a section of the network may influence payoffs in the formation of relationships in other parts of the network (among others [10–12, 20–22])). For [23] the key issues of econometric models are the issues related to their identification when the social structure represented by W is observed by the researcher, otherwise the researcher may still hope to attempt identification under plausible additional restrictions when W is partially known. The statistical approach to networking focuses on the statistical properties of joint distributions (among others [24–31]) and distinguishes between (1) simple random graph [25], (2) dyadic independent models (e.g., for directed models [26]), (3) dyadic dependence models (e.g. [29]), and (4) high-order models dependence (e.g., Pattison and Robins [32] introduced the partial conditional dependence). For a review of the statistical models of network formation, see [33]. We will consider a specification model in [26]. Where we identify the tendency of an i node to send a link to the outside α_i^{out}, α_i^{in} capturing the tendency of the j node to receive the link relative to the senderits "attractiveness", α_{rec} records the tendency for the link to be reciprocated and its high positive value will increase the probability of a symmetrical adjacence matrix (when all these parameters are zero the model corresponds to a logit where the bonds are formed in independent mode with probability equal to exp $(\alpha)/(1 + \exp(\alpha))$) (as said our statistical model inspires this). Generalizations and special cases for this model have been suggested and analyzed for example in [34, 35].Statistical literature is not (or at least not directly) related to patterns of network formation of a strategic − economic nature. In strategic models, agents form links according to an explicit notion of equilibrium and a payoff structure in which the utility function of agents $U_i(g)$ depends on the g network and the variables at the node or dyadic level. A class of these focuses on iterative formation [11, 20, 22] where at each iteration of the meeting process a couple of individuals and errors are drawn and (myopically) determine the formation or dissolution of a relationship according to the payoffs structure. Other classes of economic models focus on a static framework as in [6], in whose work they define the concept of pairwise stability for indirect networks. This last concept of equilibrium incorporates the idea that each link can be cut unilaterally but its formation requires mutual consent (as we will see, we will use this concept of equilibrium but in the case of unilateral formation of the relationship). Stable pairs networks would be staging points for the link formation sequences produced through a meeting protocol if the payoff structure does not change at each new meeting. [36] starts with the description of a simple model in which directly related agents can transfer one with the other. Therefore agents i and j will form a link if the net surplus from forming the link is positive, conditional on the behavior of forming links of other agents in the network. This is a variant of network formation game with direct transfers under the pairwise stability [37]. This rule, as stated in [36], implies that agents form links if: (1) they share friends in common and/or (2) the idiosyncratic utility unnoticed by doing so is high $(-U_{ij})$. The size of $\gamma_0 > 0$ captures the strength of the preference of agents for triadic closure in the bonds. The dependency of the surplus generated on an i-vs-j link on the presence or absence of links on the other pairs of agents constitutes an externality. Externalities generate complex interdependencies between the choices of different agents. Different rules of links formation from the one in [36] are for example in [38, 39]. In this paper we will develop an economic model that falls within static models where links are formed unilaterally [7] and simultaneously where all agents have the same payoff function, and differently to [36], we do not have externalities (we believe that this is a forced choice in a DEA context because the DEA-based network will never present transitivity and indirect links or in other words the DEA-based network efficient agents have no other links through which you can influence the usefulness of the inefficient agents related to it, or at least the direct nature of the links prevents this) and a statistical model of network formation based on DEA. The observed characteristics of the agents can be considered in the statistical model. For this network we will then define a concept of equilibrium at the level of the couple [6]. From the statistical point of view, subject to our economic model, we will propose a model of network formation with the assumption of independent direct links similar to what happens in [26]. However, unlike [26] our statistical model does not consider reciprocity effects ($\alpha_{rec} = 0$) as well as i (inefficient) does not receive links $\left(\alpha_i^{in} = 0\right)$ and j agent involved in the diade does not receive and does not send links ties ($\alpha_j^{in} = 0$ and $\alpha_j^{out} = 0$), therefore the probability of observing a direct link in the dyad i-j depends only on the effect of the agent sending. The specification of the ERGM model will therefore include independent dyadic statistics. Any analysis of the network effects on individual outcomes and related econometric issues (identification and estimation) goes beyond this paper (although a hint is offered in the training mechanism).

3   A DEA-Based Network Formation Model

In this section we develop an economic model of network formation based on DEA methodology, define a concept of equilibrium for it, clarify the formation mechanism, define the statistical model, and discuss network effects. We will assume that there are $N = 1 \ldots n$ agents who can form links between them and that the formation and deletion of a link require the consent of only one of the parties involved [7]. The ties set will form the network we denote with D, which is a binary matrix $n \times n$ (an example is the binary version of the adjacency matrix built in Appendix 2). If $d_{ij} = 1$, then i and j are connected if one of them decides to form link and, $d_{ij} = 0$ otherwise. Therefore D is an asymmetrical matrix and the relationships are only direct and not recipocated. We will normalize $D_{ii} = 0$ for all i's.

Utility Each i-esimo agent (DMU) has a $d_x \times 1$ vector of observable attributes X_i (for example proprietary form, location, size and so on) and a vector $(n - 1) \times 1$ of preferences not observable to the researcher $\varepsilon_i = (\varepsilon_{i1}, \ldots \varepsilon_{i, i-1} \ldots \varepsilon_{in})$, where ε_{ij} is the agent preference i for link ij. The utility of the agent i in a network depends on the configuration D of the network, the observable attributes X, and the unnoticed preferences ε_i,[1] for example

$$U_i\left(D, X, \varepsilon_i\right) \tag{1}$$

For each $i \cdot j$, we break D in $(d_{ij}; D_{-ij})$, where D_{-ij} is the network obtained by D after removing the link ij. Then the marginal utility of i in forming the link with j is

$$\Delta U_{ij}\left(D_{ij}, X, \varepsilon_i\right) = U_i\left(1, D_{-ij}, X, \varepsilon_i\right) - U_i\left(0, D_{-ij}, X, \varepsilon_i\right) \tag{2}$$

A possible specification of the utility function may be as follows:

$$U_i\left(D, X, \varepsilon_i\right) = \sum_{j=1}^{n} D_{ij}\left(u\left(X_{ij}\right) + \varepsilon_{ij}\right) \tag{3}$$

where X_{ij} is a $m \times 1$ vector with $m -$ emmesima component representing a quantitative measure between DMU i and j (asymmetric type), ε_{ij} is a scale collecting residual variables that will influence the net benefit that the DMU unit receives from the formation of the ij link and that are undetectable to the researcher [10]. Thus, there is no utility from indirect links [4, 11, 40]. For us it is questionable to consider in the [3] a second term that captures the additional utility for i and j from having agents in common (see Fig. 1 below). If we consider the marginal utility of i from forming the link with j, which is given by

$$\Delta U_{ij}\left(D_{-ij}, X_{ij}, \varepsilon_{ij}\right) = u\left(X_{ij}, \beta\right) \tag{4}$$

Then it consists only of the direct utility to j. This specification differs from that of [22, 40] where externalities are considered through indirect links.

Equilibrium We assume that agents observe X and ε, so that it is a complete information game. The equilibrium concept we will consider is pairwise stability

Fig. 1 Common agent

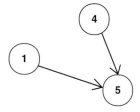

---

[1]McFadden and Manski [60, 61] modeled a discrete choice problem for a single agent using a random utility model (RUMs). These models offer a way to infer the distribution of preferences from the observed distribution of choices. Unfortunately, as is familiar from the literature on game theory, the choice of agents is interdependent, as can be the case with network formation.

[62] for non-transferable utility (NTU) [6] although here the DEA-based network is similar to a one-flow model [7], so we will propose to define the concept of pairwise stability for the DEA-based network as follows:

Definition 1 DEA-based network will have pairwise stability if

- For each $D_{\mathrm{DEA}ij}=1$, $\Delta U_{ij}(D_{\mathrm{DEA}-ij}, X_i, X_j, \varepsilon_{ij}) > 0$ and $\Delta U_{ji}(D_{\mathrm{DEA}-ji}, X_j, X_i, \varepsilon_{ji}) = 0$;
- For each $D_{\mathrm{DEA}ij} = 0$, $\Delta U_{ij}(D_{\mathrm{DEA}-ij}, X_i, X_j, \varepsilon_{ij}) = 0 \Longrightarrow \Delta U_{ji}(D_{\mathrm{DEA}-ji}, X_j, X_i, \varepsilon_{ji}) = 0$.

So, a possible choice can be for example

$$D_{\mathrm{DEA}ij}=1\left\{\Delta U_{ij}\left(D_{\mathrm{DEA}-ij}, X_i, X_j, \varepsilon_{ij}\right) >0, \Delta U_{ji}\left(D_{\mathrm{DEA}ij}, X_j, X_i\varepsilon_{ji}\right) =0\right\} \forall i \neq j \tag{5}$$

and assuming $\varepsilon_{ij} = 0$ we will have

$$D_{\mathrm{DEA}ij} = 1\left\{\Delta U_{ij}\left(D_{\mathrm{DEA}-ij}, X_{i,}X_j\right) > 0, \Delta U_{ji}\left(D_{\mathrm{DEA}ij}, X_j, X_i\right) = 0\right\} \forall i \neq j \tag{5a}$$

The choice of a $D_{\mathrm{DEA}ij}$ link in (5) depends on the choice of other $D_{\mathrm{DEA}-ij}$. This indicates that we cannot treat each linkas a single observation and use a dyadic regression because $D_{\mathrm{DEA}ij}$ is endogenous in the model because it can be related to $(\varepsilon_{ij}, \varepsilon_{ji})$. But since $\Delta U_{ji} = 0$ the choice is only for the agent i with ε_{ij}. What futher complicates the statistical inference of (5) is that there are multiple equilibrium, which will influence the identification of the parameters (but this is not done here and may be of interest for further work developments). In (5a), however, we can assume independence of choice (because $\varepsilon_{ij} = 0$). At the same time we do not have reciprocity because ΔU_{ji} (the marginal utility of the agent j in the link ij) will always be zero and the agent j will have no incentive to form some kind of link with i.

Network Formation Mechanism (and a Hint at Econometric Specification) DMUs simultaneously announce the desired output link (we remind that the DMU i-th unilaterally decides on the formation of the link ij according to the indications of the analysis DEA of peers) and this happens under the assumption of complete information. So each DMU i gets a payoff as in (3) under the pairwise stability in (5a). Once the mechanism is specified the D_{DEA} network is a strategy in Nash equilibrium if it resolves the following system of $N(N-1)$ equations:

$$D_{\mathrm{DEA}ij} = 1\left\{\theta X_{ij} + \varepsilon_{ij} \geq 0\right\} \quad \forall i \neq j \text{ with } i, j \in D_{\mathrm{DEA}ij} \tag{6}$$

A graphical example of simultaneity is shown in Fig. 2.

The inefficient agents with $U > 0$ will announce the desired output link according to the strategy defined by the peer group analysis as developed in Appendix 1. For example following the example of Appendix 1 the unit C will have a strategy training of two links with agents E and F. In summary we will say that "in this DEA-

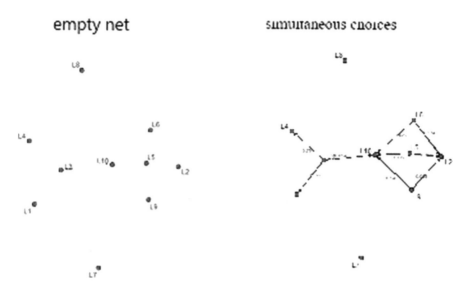

Fig. 2 Simultaneous mechanism formation of DEA-based network

Table 1 Distribution of net utility under $c_{ij} = 0$

DMU	Formula (7)	
1	$(1 - 1) \times 0 = 0$	–
2	$(1 - 0.858) \times 0.322 = 0.0457$	$(1 - 0.858) \times 0.677 = 0.096$
3	$(1 - 0.789) \times 0.434 = 0.092$	$(1 - 0.789) \times 0.565 = 0.119$
4	$(1 - 0.828) \times 0.596 = 0.102$	$(1 - 0.8281) \times 0.403 = 0.069$
5	$(1 - 1) \times 0 = 0$	
6	$(1 - 1) \times 0 = 0$	

based network version agents will strategically form links with other agents if the marginal utility is positive $\Delta U > 0$ and with probability of $\lambda^{\cdot}_{ij}$ independently of other links" [36] shall apply.

Example Suppose that there are $N = 6$ DMUs with the same technology (2 inputs and 1 output) and suppose that once applied the DEA (model (11) in Appendix 1) the DEA score efficiency for each of it are the following: 1, 0.858, 0.789, 0.828, 1, 1 (at this time the network is empty, yet). We now assume a further specification of the utility function (3) (with $\varepsilon_{ij} = 0$) as follows:

$$U_i = \sum_j^{1-N} (1 - \text{DEAeffscore}) \, \lambda^{\cdot}_{ij} - c_{ij} \qquad (7)$$

where $\lambda^{\cdot}_{ij}$ are the optimal lambda and c_{ij} i costs to maintain and form links. Suppose we then have a lambda matrix like the one in Appendix 2. Assuming $c_{ij} = 0$ we will have the following net utility distribution (Table 1):

In the empty network we will have $U_i = 0$, always. The marginal utility ΔU for DMU 3 will be distributed for example as follows: $3 \rightarrow 5{:}0.092$ and $3 \rightarrow 6{:}0.119$. That is, unit 3 will have a marginal utility of 0.119 in forming the link with unit 6 and a marginal utility of 0.092 in forming the link with unit 5 (so that the sum will be just equal to $1{-}0.789 = 0.211$ $(0.092 + 0.119)$). In the end, all strategies have been completed (simultaneously) resulting in the same network as right picture in Fig. 2.

Statistical Point of View Conditional on (5a) the DEA-based network is based on independent choices to only send links to the outside. Statistically speaking this allows us to focus on $D_{ij} = (X_{ij})$ dyads with only these states : (1) $D_{ij} = 0$ dyad null (if $U = 0$ for the agent i and $U = 0$ for agent j), and (2) $D_{ij} = 1$ dyad asymmetric $(i{\cdot}j)$ (if $U > 0$ for agent i and $U = 0$ for agent j). In the case of reciprocity it would mean that the efficient agent also sends a link to the inefficient one. But this would contradict the above economic model as it would contradict the "lesson of learning from the best" by the inefficient agent. The link resulting from the agent's strategic decision can be considered as a random variable with some type of distribution and therefore the adjacency matrix in Appendix 2 considered as a possible realization of this random variable shall be used. Under the assumption of independence the links between the nodes follow a distribution of Bernoulli with equal probability, say p. And as we know for $n\rightarrow$ is a small p the distribution of degrees will follow a Poisson and then we will have a random graph model [25, 41]. However this is not our case where by placing p_{ij} $(1,0)$, the probability of sending a link by agent i to the agent j, and p_{ij} $(0, 0)$, the probability of null dyad following [26], we can write the following simple statistical model:

$$\log p_{ij}\ (0,0) = \lambda_{ij} \quad \text{and} \quad \log p_{ij}\ (1,0) = \lambda_{ij} + \alpha_i + \theta \tag{8}$$

$\lambda_{ij}, \alpha_i, \theta$ are the parameters of the model. α_i controls the effect of a link leaving i, θ is the general propensity of the DEA network to have a link and λ_{ij} is a normalization constant that ensures (2) to maintain for each dyad (i, j).

$$p_{ij}\ (0, 0) + p_{ij}\ (1, 0) = 1 \tag{9}$$

Thus the model (8) does not include any in-link effects or/and reciprocity. The outgoing link is now strategically based.

Network Effects Modeling the DEA-based network under the assumption of independence means that the agent's behavior does not depend on the relationships and actions of others in the rest of the network. In other words from this assumption follows the absence of network effects. Assuming that each individual has preference given by (7) the utility of agents when they take action is $(1 - \text{DEAeffscore}) * \lambda_{ij}$ and there is the cost of this action. Thus agents take actions $\alpha_i = 1$ if $(1 - \text{DEAeffscore}) \lambda_{ij} > c$ and do not take actions $\alpha_i = 0$ if $(1 - \text{DEAeffscore}) \lambda_{ij} \leq c$. Assuming that $c = 0$ means that in the absence of network effects a proportion of agents for which $(1 - \text{DEAeffscore}) \lambda_{ij} > 0$ will take action for which DEAeffscore* $\lambda_{ij} < 1$. In other words equilibrium in the absence of network effects only inefficient agents will take action.

4 Micro and Macro Analysis

Suppose you have a sample of $N = 40$ units using the same technology (two inputs and an output) and have a DEA-based network. The questions for a researcher can be: (1) the network is dense? (2) how many connections exist on average in the network? (3) is there any unit that forms more ties than the others? the links have the same structures? And so on.

4.1 Macro Analysis

Our macro analysis will consider: (1) network density, (2) average outdegree, and (3) average distance and diameter. In simple patterns of network formation, the distribution of Poisson emerges when the links are formed uniformly at random (and are also dense) so that the difference in grade between nodes reflects causality relative to a binomial variable. A distribution exhibiting the law of power instead has a large variation in degrees and is usually derived from a process "rich seeks rich": a dynamic in which the nodes with a higher grade are the nodes gaining new ties at a higher rate [42, 43]. Differences in grade distribution have an important implication for diffusion processes. The average degree between nodes equals the number of ties along the shortest route between nodes. The average distance between them has important consequences from the economic point of view. The fact that social networks have an average short distance is the characteristic studied by Milgram [44]. Generally route lengths tend to vary with the logarithm of the number of nodes in a network. Theorems that have presented that this is true have been tested for many network models, starting with some simple ones such as [25, 45]. The diameters tend to be small in random graphs for the reasons mentioned above, while it is for many different reasons that the diameter is short for networks in which individuals choose links strategically. The diameters are directly useful to offer limits for some diffusion processes that travel through shorter paths [46].

As we can see our network of 40 DEA-based virtual enterprises forms just 6% of all possible links (this measure considers all possible links without distinguishing between in and out), and on average each sender (agent proposing the link) sends 2 proposals for links (see Table 2). Reaching every other agent is relatively easy (average diameter is 1). Very interesting in this is that our network is the distribution of the out and in degree. The second is very concentrated with a coefficient of Gini equal to 0.953 (see Table 2), while the first is relatively uniform with a coefficient of Gini equal to 0.211 (see also the Lorenz curves in Fig. 1). Our interpretation of Gini's coefficient for the outdegree, in this example, is that the probability for a node

Table 2 Macro level analysis

Measures	Value
Density	0.058
Average outdegree	2.275
Average indegree	2.275
Average diameter	1
Gini outdegree coefficient	0.211
Gini indegree coefficient	0.953
Efficiency	0.966

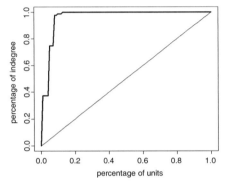
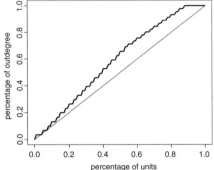

Fig. 3 Lorenz curve for in/out degree

Table 3 Triad census

Type of triad	Frequency
003	7831
012	640
102	0
021D	77
021U	1332
All others	0

to have some outdegree is about the same (under the assumption of independent choices).

The Gini coefficient for indegree confirms the concentration of sender choices on a few items as we can see in Fig. 3. Our macro analysis concludes with structural analyses and may concern the census of the dyads and the triads, here we will report only the analysis of the triads as in Table 3.

Triad 003 means that there are no mutual links (mutuality $= 0$), that there are no asymmetric links (asymmetry $= 0$), but that there are three null dyads in the triad (nullity $= 3$). The interpretations of the triads in the case of DEA-based network are, in our opinion, the following: (1) 003 means empty triad containing agents of the same type (or all efficient or all inefficient) or a combination of them, (2) 012 means that the triad contain one asymmetrical link and two null dyads. This means that in the DEA-based network this triad can contain one inefficient and two efficient agents, only, (3) 021 U means that the triad contain two asymmetrical dyads and one null dyad. This means that it can contain two inefficient agents and one efficient agent only, (4) 021D means triad can contain two asymmetric dyads and one null dyad. This triad is similar to the last one (i.e. 021U) but can contain one inefficient agent and two efficient ones, only finally (5) we observe the triad 021C is illogical in a DEA-based network as this triad would assume the existence of a link from the efficient agent to the inefficient one, that is a $j \rightarrow i$ link. All remaining triads with 1 and 2 as the first element (for example 111D, or 201, and so on) cannot (and should not) exist in a DEA-based network because they presuppose a mutual link of the type $i \rightarrow j$ and $j \rightarrow j$. As well as triads of type 030T and 030C assuming transitivity or cycles in them cannot exist in a network just on the DEA.

4.2 Micro Analysis

Micro level analysis concerns the centrality and structural equivalence. There are numerous measures of centrality [47–49] and they can affect concepts such as influence, prestige, popularity, and so on [50, 51]. These last measures have implications for the processes of diffusion within networks, or the impact of a node on transactions [46]. Two nodes are structural equivalents if they have the same relationship with all other nodes [49, 52, 53]. The phenomena which have taken this measure into account include the evolution of inter-organizational networks (i.e., [54]) or the diffusion of innovation (i.e., [55]) (Table 4).

For the DEA-based network we will say that two nodes are structurally equivalent if they send the same links to the same efficient agents (see Fig. 4 below).

Table 4 Indegree of DEA-based network of 40 virtual firms

DEA full efficient actor	Indegree
8	34
13	34
21	21
24	1
34	1

Fig. 4 Structural equivalence example

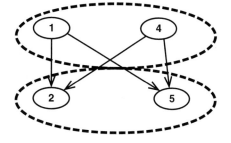

5 Estimation and Simulation

As presented above we are assuming our statistical model (8) belongs to the family of exponential random graph models [26]. That said, we place the following log-linear specification:

$$\log\left[\mathbb{P}_{\theta}, y\left(Y = y_{ij}\right)\right] = \sum_{i \neq j}\sum \phi_{ij} y_{ij} - \log \kappa\left(\theta, y\right) \tag{10}$$

where $\phi_{ij} = \alpha_i + \theta$ and α_i are the sender's effect and the general effect of the links similar to the interception of a linear regression. The specification in (10) differs from the model p_1 [26] because no transitivity effect is considered and the term ϕ_{ij} considers only the effect of the sender. The (10) model can be estimated using the estimation method of approximate maximum likelihood estimates obtained using a stochastic algorithm based on the Markov Chain Monte Carlo (MCMC). Obviously, in our intention model (10) should generate a network with the same structural proprieties of a DEA-based network. The estimates are given in Table 5 below.

From Table 5 we observe a positive sender effect for inefficient agents and a negative effect for efficient ones (8,13,21,24,34 are DEA efficient nodes). The expected probability of observing a direct link from the inefficient agent sender in the DEA network to the efficient one in the current network is increasing and statistically significant and the expected probability of observing a direct link from the efficient agent never sender in the DEA-based network to the inefficient one is decreasing and statistically significant.

5.1 Simulation

In this section: (1) we will simulate different statistical models assumed as benchmarks and compare them with our network, (2) we simulate DEA-based networks of different sizes and (3) we simulate net from our ERGM model and compare it

Table 5 ERGM estimates

	Estimates	Std. error	Pr $(>|z|)$
Sender (2,4,5,6,9,11,20,22,25,27,30,32,35)	6.0404	1.1713	<1e^{-04}***
Sender (3,7,10,12,14,15,16,17,18,19,23,26,28,29,36,37, 38,39,40)	6.8186	1.1689	<1e^{-04}***
Sender (8,13,21,24,34)	−INF	0	<1e^{-04}***
Sender (33)	1.1689	1.1689	<1e^{-04}***
Node.Factor = Status.Ineff	−6.4743	0.7262	<1e^{-04}***
AIC	609.2	BIC	823.3

*** Significance at 0.001%

Table 6 Triad census of simulated net

Type of triad	Bernoulli $(p = \lambda_{ij}^*)$	Bernoulli $(p = 0.008)$	Erdos–Renyi $(e = 91)$	Erdos–Renyi $(p = 0.008)$	Uniform MAN distribution $(m = 0, a = 91, n = 689)$	Barabasi and Albert "preferential attachment" (power = 1)
	Frequency					
003	8940	9321	6886	9428	6980	8496
012	816	548	2548	448	2583	2186
102	0	0	66	0	0	0
021D	2	3	89	0	64	0
021U	122	1	100	2	73	64
021C	0	7	173	2	167	34
111D	0	0	3	0	0	0
111 U	0	0	7	0	0	0
030 T	0	0	6	0	9	0
030C	0	0	2	0	4	0
All others	0	0	0	0	0	0

Table 7 Outdegree Gini coefficient

Measure	Bernoulli $(p = \lambda_{ij})$	Bernoulli $(p = 0.008)$	Erdos–Renyi $(e = 91)$	Erdos–Renyi $(p = 0.008)$	UMAN $(m = 0, a = 91, n = 689)$	Barabasi "preferential attachment"
Outdegree Gini coefficient	0.40293	0.663	0.3755	0.715	0.282051	0.025641

with ours. The first benchmark model is random graph model [25] which for n, D can be approximated with a causal variable of Poisson $\lambda = (n \cdot 1) p$. The second is "Preferential attachment" model [42]. Table 6 below reports the triad analysis results to be compared with Table 3.

The Gini coefficient for the outdegree is in Table 7. The UMAN network [27] is conditioned with: Mutual = 0, asymmetric = 91, and null dyads = 689. For Erdos–Renyi the condition on asymmetrical ties. In the Barabasi–Albert we place power = 1 (linear preferential attachment). From the structural point of view the Bernoulli network is more similar to our DEA-based network, although the distribution of the outdegree is less uniform (Gini = 0.40293 and 0.663 in Table 7) than that of the DEA network in our example (Gini = 0.211 in Table 2). This would be partially of comfort for our statistical model. In Fig. 5 we report the graph of the densities of the outdegree for DEA networks of different sizes: $N = 40$, 90, 250, and 500 and the densities of 3 theoretical distributions (uniform, Bernoulli). In Table 8 and Fig. 6 we report respectively the Gini coefficient of the outdegree, with other statistics, for the DEA network and the outdegree Lorenz curves, respectively.

A zero skewness means a distribution centered on the average, while a value for this statistic 0 means a queue moved to the left while a value >0 means a queue

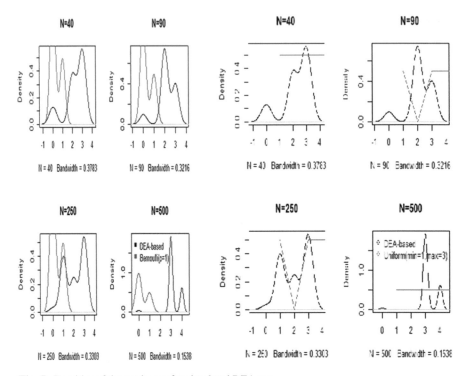

Fig. 5 Densities of the outdegree for simulated DEA net

Table 8 Outdegree Gini coefficient for simulated DEA net

Measure	$N = 40$	$N = 90$	$N = 250$	$N = 500$
Outdegree Gini coefficient	0.2110	0.1688	0.2453	0.0727
Mean	2.275	2.275	2.275	2.275
s.d.	0.9867	0.7825	0.9401	0.5921
Skewness	-1.13311	-1.267	-0.3209	-1.928

moved to the right. In our DEA-based network we observe a negative value for this statistic, this means some high values and relatively low values compared to the average (or differently said the median of the outdegree is less than the average of it or that there are outliers farther from the average). So, some ideas (including dynamics) on the formation of this network can be summarized as follows: as the number of agents increases each new agent tends to send on average the same number of links but there are more and more agents who send a lower than average number of links (a higher negative skewness value for $N = 500$ in Table 8 and the distribution of the outdegree is more uniform) (Gini coefficient for the outdegree near r zero) or, we can say that the number of agents sending 1.2. n ties is about the same in each group). We now report triad analysis for the ERGM models estimated above (Table 9).

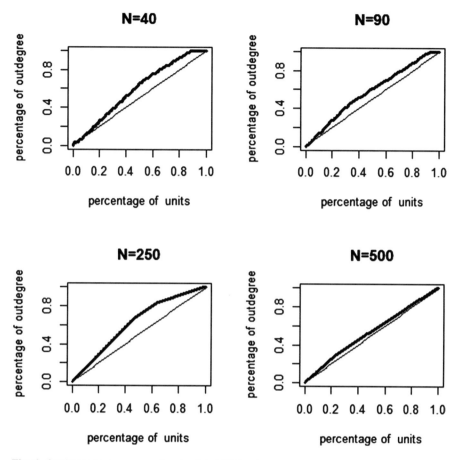

Fig. 6 Outdegree Lorenz curves for simulated DEA net

All networks simulated by the estimated ERGM model here are structurally different from the DEA-based network (Tables 3, 6, and 9). This suggests a few things: (1) or the ERGM model is not well specified , (2) you need to introduce some dependency, or (3) you need to check for a degeneration problem. One thing we do first is constrain our ERGM model on the distribution of grades according to the observed network. The result of this choice is represented by the analysis of the triads of the models simulated by this new ERGM model (Table 9). Once controlled for the distribution of degrees, the simulation generates networks more similar to our DEA network than in the absence of constraints. These constraints replicate the distribution of outgoing ties grades observed in our application (on average 2.75). The simulation shows that the statistical model in (8) therefore (10) generates networks structurally similar to our network based on DEA but is not the same when it generates a type of transitive triad (030T) which is absent from the DEA-based network structure. The difference in structure between the simulated networks

Table 9 Triad census of ERGM models

Type of triad	Ergm sim 1	Ergm sim 2	Ergm sim 3	Ergm sim 4	Ergm sim 5	Ergm sim 6	Ergm sim 7	Ergm sim8
	Frequency							
003	184	163	108	128	152	167	116	112
012	1042	858	829	826	874	885	763	791
102	493	418	329	422	481	429	338	413
021D	512	480	501	444	461	458	435	447
021U	527	450	528	479	467	517	462	443
021C	980	987	1063	929	895	950	856	870
111D	927	938	835	889	1001	875	900	927
111U	897	902	822	923	878	945	874	916
030T	1021	1088	1280	987	888	949	1075	953
030C	333	348	423	320	281	333	350	295
201	401	384	299	448	466	422	421	493
120D	406	475	471	486	525	472	544	523
120U	429	471	499	463	465	443	531	478
120C	856	923	1008	944	923	971	1057	1037
210	776	850	795	1037	936	913	998	998
300	96	145	90	155	187	151	160	184
Constrained ERGM								
003	7831	7832	7833	7833	7832	7833	7833	7833
012	642	641	639	640	641	642	639	644
102	0	0	0	0	0	0	0	0
021D	75	74	73	74	73	74	73	73
021U	1330	1329	1329	1328	1329	1328	1328	1328
021C	0	1	2	2	1	2	2	0
111D	0	0	0	0	0	0	0	0
111U	0	0	0	0	0	0	0	0
030T	2	3	3	4	3	4	4	4
030C	0	0	0	0	0	0	0	0
201	0	0	0	0	0	0	0	0
120D	0	0	0	0	0	0	0	0
120U	0	0	0	0	0	0	0	0
120C	0	0	0	0	0	0	0	0
210	0	0	0	0	0	0	0	0
300	0	0	0	0	0	0	0	0

from ERGM model and the empirical DEA-based network (those represented of the optimal lambda adjacency matrix) is, according us, related to the assumption of independence in our statistical model. Therefore a possible future line of research can follow the direction of introducing some kind of addiction, obviously both for the statistical model and for the strategic economic one. This narrows the attention on the introduction of some type of (structural) dependence in our statistical model.

6 Discussion and Conclusions

In this paper, we have considered the formation of a DEA-based network by proposing both a strategic economic model and a statistical model based on a previous study. Our economic model assumes that agents form links if their marginal utility from "not forming the link" to "forming the link" is positive. In addition, the other agents involved in the bond receive zero marginal utility. This is because, in the proposed model, the agents who receive the invitation to form the bond are entirely efficient agents who have no incentive to form bonds (i.e., $\Delta U = 0$). Both the formation and the eventual cancellation of the bond is unilaterally proposed by the inefficient agent (refer [7]). The statistical formulation of the network formation model based on the DEA replicates the strategic behavior of the agent that proposes the link. Therefore, the statistical model does not consider either the interdependence of the decisions or the possibility of reciprocal or transitivity in the bonds. This means that the statistical model (proposed in Sect. 3) considers only two stages of the dyad: (1) null (no link is sent) and (2) asymmetric (only the agent concerned sends the link) [26]. The lack of interdependence is a limitation of our modeling (both economic and statistical). In order to construct a DEA model the constraint system should be introduced first. However, this offers the opportunity for further research. The statistical model is the basis for the specification and estimation of an ERGM model that considers only the sending effect. The simulations show that the specification of the statistical model generates networks that are structurally similar but not equivalent to a DEA-based network (and this can be seen by comparing the analyses of the triads in Tables 3 and 9). This points to the incomplete search for the statistical model and the specification of a more appropriate ERGM model. With some probability, the introduction of dyadic dependency is the future line of research in the nearest time. Meanwhile, our statistical model is based on that proposed in a previous study [26] but differs with respect to the lack of reciprocity and the lack of attractiveness. From the estimation point of view, it can be assumed that the most appropriate distribution is the exponential one. However, a possible advantage of having assumed independence at the level of dyads is that we can reflect the spirit of "learning from the best" in a competitive context that characterizes the DEA approach [56]. In our opinion, the paper contains several innovative elements with respect to the reference literature: (1) the use of DEA methodology as a generator of relational variables, (2) the definition of the couple equilibrium for the DEA-based network, and (3) the strategy agents are based on the optimal solutions of a mathematical program. The first condition in Definition 1 requires that every link present in a stable DEA network must be profitable for a single agent. In particular, it is profitable for the agent who has an interest in improving its efficiency. The second condition requires that in the pairwise stable network in the absence of a link the marginal utility for the sending agent is zero and for the other there is no incentive to create a link. In the DEA network, this last is partially different for the efficient agent for whom it is not profitable to act. In conclusion, although the paper offers an initial starting point for further research, especially as regards the statistical model, it offers an interesting and new framework for network formation models. All calculations and analyses were conducted in R [57].

Acknowledgments I would like to thank all participants for their useful comments during my seminar held at the Department of Statistics and Economics Science of the University of Salerno in the date 9 May 2018 where I showed a preliminary version of this work and the ARS'17 participants for their useful comments. Yet I would like to thank editor and anonymous referees for detailed comments on earlier versions of the paper.

A.1 Appendix 1 DEA Analysis of Peers

DEA [58, 59] is a technique used to measure the relative performance of different types of entities (called decision making units DMU). In it the best receive a score always equal to 1 while the remaining a score less than 1. As seen from Fig. 7.

As can be seen from this example figure, point B is an inefficient point that to become efficient must change its technology into a new B' technology that coincides with the convex combination of the technologies of units C and D on the border. This new technology does not exist but is the result of a mathematical program. According to [56] the DEA methodology incorporate the lesson "learning from the best": B (inefficient) from C and D (both full efficient) in Fig. 7. The DEA peer group analysis can be runned as follow:

1. Solve the following mathematical program:

$$\min \theta$$
$$s.t. \quad \sum_{i=1}^{N} \lambda_i x_i \leq \theta x_{io} \tag{11}$$
$$\sum_{i}^{N} \lambda_i y_i \geq y_{io}$$

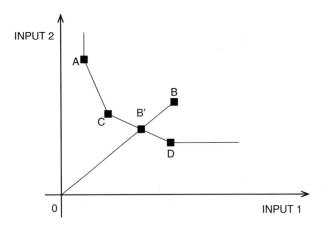

Fig. 7 DEA frontier

2. Consider the left side of the constraints of 1 and define the reference units as follows:

$$\left(\sum_{n=1}^{N} \lambda^n x^n, \sum_{n=1}^{N} \lambda^n y^n\right) \tag{12}$$

All units with very good positive lambda values at the best have their peers. To clarify, if for example we have $n = 6$ sample observations of entities that do a certain thing using $m = 2$ inputs and $p = 1$ outputs and apply the following program (which is the computational version of the above):

$$\min \theta$$
$$s.t. \ \theta x 40 \geq \lambda^A x 80 + \lambda^B x 90 + \lambda^C x 90 + \lambda^D x 75 + \lambda^E x 40 + \lambda^F x 95$$
$$\theta x 70 \geq \lambda^A x 45 + \lambda^B x 50 + \lambda^C x 60 + \lambda^D x 65 + \lambda^E x 70 + \lambda^F x 30 \tag{13}$$
$$30 \leq \lambda^A x 40 + \lambda^B x 30 + \lambda^C x 10 + \lambda^D x 20 + \lambda^E x 30 + \lambda^F x 40$$

we get, always by way of example, the following excellent solutions (θ^*, λ (A^*), λ (D^*)) for example for a unit E we have λ (A^*) = 0.35 and λ (D^*) = 0.75 this will have the same unit D (point B in Fig. 7) would then be 35% and 65% respectively of unit D. In this way the unit E becomes fully efficient $\theta = 1$. This step, as will be evident in Appendix 2 below, is crucial as it provides excellent lambda values.

B.1   Appendix 2 Adjacency Matrix for DEA-Based Network Representation

The optimal DEA solutions (11) provide us with the optimal lambda λ values for each unit. Therefore once the peer matrix is formed (see Table 10 below) this matrix can be transformed into an adjacency matrix (see Table 11 below) to represent a network, transforming the row and column of headers into actors who send (the top column) and actors who receive (the first line).

Table 10 Peers matrix

Units	Peers					
	A	B	C	D	E	F
A	1	0	0	0	0	0
B	0	0	0	0	0.322	0.677
C	0	0	0	0	0.434	0.565
D	0	0	0	1	0.596	0.403
E	0	0	0	0	1	0
F	0	0	0	0	0	1

Table 11 Adjacency matrix
for DEA net representation

Units	A	B	C	D	E	F
A	1	0	0	0	0	0
B	0	0	0	0	0.322	0.677
C	0	0	0	0	0.434	0.565
D	0	0	0	0	0.596	0.403
E	0	0	0	0	1	0
F	0	0	0	0	0	1

In this way the lambda become relationship variables while its values can receive different interpretations (for example a probability). We in the text used the binary version of this matrix.

It is important to note that the value 1, when indicating a fully efficient unit, should not be confused in the peer matrix and the value 1 when indicating a peer to be fully imitated. In the first case, in the passage from the matrix of the pairs to the matrix of adjacence the first 1 would be a loop (relation with itself from part of the efficient unit) while the second one a relation with another unit. From the adjacency matrix, which is a weighed matrix it should be noted that: (1) some actors (the efficient ones) can only receive, (2) that there are always no links, (3) and that there will never be observed links between efficient units. These points can be stylized "factors" for our DEA network.

References

1. Calvó-Armengol, A., Jackson, M.: The effects of social networks on employment and inequality. Am. Econ. Rev. **94**(3), 426–454 (2004)
2. Calvó-Armengola, A., Jackson, M.: Networks in labor markets: wage and employment dynamics and inequality. J. Econ. Theory. **132**(1), 27–46 (2007)
3. Calvò-Armengol, A.: Job contact networks. J. Econ. Theory. **115**(1), 191–206 (2004)
4. Calvo-Armengol, A., Patacchini, E., Zenou, Y.: Peer effects and social networks in education. Rev. Econ. Stud. **76**(4), 1239–1267 (2009)
5. Conley, T., Udry, C.: Learning about a new technology: Pineapple in Ghana. Am. Econ. Rev. **100**(1), 35–69 (2010)
6. Jackson, M., Wolinsky, A.: A strategic model of social and economic networks. J. Econ. Theory. **71**, 44–74 (1996)
7. Bala, V., Goyal, S.: A noncooperative model of network formation. Econometrica. **68**(5), 1181–1229 (2000)
8. Jackson, M.: An overview of social networks and economic applications. In: Handbook of social economics. Elsevier, Amsterdam (2011)
9. Jackson, M.: A Survey of Models of Network Formation: Stability and Efficiency. California Institute of Technology, Pasadena (2003)
10. Fafchamps, M., Gubert, F.: The formation of risk sharing networks. J. Dev. Econ. **83**(2), 326–350 (2007)
11. Mele, A.: A structural model of dense network formation. Econometrica. **85**(3), 825–850 (2017)

12. Sheng, S.: A structural econometric analysis of network formation games, UCLA Working Paper (2016)
13. Bajari, P., Hong, H., Ryan, S.: Identification and estimation of a discrete game of complete information. Econometrica. **78**(5), 1529–1568 (2010)
14. Leung, M.: Two-step estimation of network formation models with incomplete information. J. Econ. **188**(1), 182–195 (2015)
15. Tamer, E.: Incomplete simultaneous discrete response model with multiple equilibria. Rev. Econ. Stud. **70**(1), 147–167 (2003)
16. Myerson, R.: Game Theory: Analysis of Conflict. Harvard University Press, Cambridge (1991)
17. Calvò-Armengol, A., Ilkilic, R.: Pairwise-stability and Nash equilibria in network formation. Int. J. Game Theory. **38**, 51–79 (2009)
18. Chandrasekhar, A.: Econometrics of network formation. In: The Oxford Handbook of the Economics of Networks, pp. 303–357. Oxford University Press, New York (2016)
19. Manski, C.F.: Identification of endogenous social effects: the reflection problems. Rev. Econ. Stud. **60**(3), 531–542 (1993)
20. Badev, A.: Discrete games in endogenous networks: theory and policy. Dissertation (2013)
21. Currarini, S., Jackson, M., Pin, P.: An economic model of friendship: homophily, minorities, and segregation. Econometrica. **77**(4), 1003–1045 (2009)
22. Christakis, N., Fowler, J., Imbens, G., Kalyanaraman, K.: An Empirical Model for Strategic Network Formation. Harvard University, Cambridge (2010)
23. De Paula, A.: Econometrics of network models. In: Advances in Economics and Econometric, Eleventh World Congress, pp. 268–323. Cambridge University Press, Cambridge (2017)
24. Morris, M., Handcock, M., Hunter, D.: Specification of exponential family random graph models: terms and computational aspects. J. Stat. Softw. **24**, 1548–7660 (2008)
25. Erdos, P., Renyi, A.: On random graphs. Publ. Math. **6**, 290–297 (1959)
26. Holland, P., Leinhardt, S.: An exponential family of probability distributions for directed graphs. J. Am. Stat. Assoc. **76**(373), 33–50 (1981)
27. Holland, P., Leinhad, S.: The statistical analysis of local structure in social networks, NBER Working Paper No. w0044 (1974)
28. Albert, R., Barabasi, A.: Statistical mechanics of complex networks. Rev. Mod. Phys. **74**(1), 47–97 (2002)
29. Frank, O., Strauss, D.: Markov graph. J. Am. Stat. Assoc. **8**, 832–842 (1986)
30. Strauss, D., Ikeda, M.: Pseudolikelihood estimation for social networks. J. Am. Stat. Assoc. **85**, 204–212 (1990)
31. Handcock, M.: Statistical models for social networks: inference and degeneracy. In: Dynamic Social Network Modeling and Analysis, vol. 126, pp. 229–252. The National Academies Press, Washington (2003)
32. Pattison, P., Robins, G.: Neighborhood-based models for social networks. Sociol. Methodol. **32**(1), 301–337 (2002)
33. Goldenberg, A., Zheng, A., Fienberg, S., Airoldi, E.: A survey of statistical network models. Found. Trends Learn. Mach. **2**(1), 129–123 (2010)
34. Dzemski, A.: An Empirical Model of Dyadic Link Formation in a Network with Unobserved Heterogeneity. University of Manheim, Mannheim (2014)
35. Hoff, P.: Bilinear mixed-effects models for dyadic data. J. Am. Stat. Assoc. **100**, 286–295 (2005)
36. Graham, B.: Methods of identification in social networks. Annu. Rev. Econ. **7**, 465–485 (2015)
37. Bloch, F., Jackson, M.: The formation of networks with transfers among players. J. Econ. Theory. **113**(1), 83–110 (2007)
38. Graham, B.: An empirical model of network formation: detecting homophily when agents are heterogeneous, NBER w20341 (2014)
39. Handcock, M., Raftery, A., Tantrum, J.: Model-based clustering for social networks. J. R. Stat. Soc. A. Stat. Soc. **170**, 301–354 (2007)
40. Mele, A.: A structural model of segregation in social networks. Centre for Microdata Methods and Practice, London (2013)

41. Gilbert, E.: Random graphs. Ann. Math. Stat. **30**(4), 1141–1144 (1959)
42. Barabasi, A., Albert, R.: Emergence of scaling in random networks. Science. **286**(5439), 509–512 (1999)
43. Jackson, M., Rogers, B.: Meeting strangers and friends of friends: how random are social networks? Am. Econ. Rev. **97**(3), 890–915 (2007)
44. Milgram, S.: The small-world problem. Psychol. Today. **1**(1), 60–67 (1967)
45. Watts, D., Strogatzm, S.: Collective dynamics of 'small-world' networks. Nature. **393**, 440–442 (1998)
46. Jackson, M., Rogers, O., Brian, W., Zenou, Y.: The Economic Consequences of Social Network Structure. Centre for Economic Policy Research., London (2016)
47. Freeman, L.: Centrality in social networks: conceptual clarification. Soc. Networks. **1**(3), 215–239 (1979)
48. Jackson, M.: Social and economic networks. Princeton University Press, Princeton (2008)
49. Wasserman, S., Faust, K.: Social Network Analysis: Methods and Applications. Cambridge University Press, New York (1994)
50. Gould, R., Fernandez, R.: Structures of mediation: a formal approach to brokerage in transaction networks. Sociol. Methodol. **19**, 89–126 (1989)
51. Goyala, S., Vega-Redondoa, F.: Structural holes in social networks. J. Econ. Theory. **137**, 460–492 (2007)
52. Burt, R.: Positions in networks. Soc. Forces. **55**, 93–122 (1976)
53. Lorrain, F., White, H.: Structural equivalence of individuals in social networks. J. Math. Sociol. **1**, 49–80 (1971)
54. Walker, G., Kogut, B., Weijian, S.: Social capital, structural holes and the formation of an industry network. Organ. Sci. **8**(2), 109–125 (1997)
55. Burt, R.: Social contagion and innovation: cohesion versus structural equivalence. Am. J. Sociol. **92**, 1287–1335 (1987)
56. Bogetof, P., Otto, L.: Benchmarking with DEA, SFA, and R. Springer, New York (2011)
57. RCoreTeam, R: A language and environment for statistical computing, Vienna (2016)
58. Charnes, A., Cooper, W.W., Rhodes, E.: Measuring the efficiency of decision making units. Eur. J. Oper. Res. **2**, 429–444 (1978)
59. Cooper, W., Seiford, L., Zhu, J., Tone, K.: Some models and measures for evaluating performances with DEA: past accomplishments and future prospects. J. Prod. Anal. **28**, 151–163 (2007)
60. McFadden, D.L.: Conditional logit analysis of qualitative choice behavior. In: Frontiers in Econometrics, pp. 105–142. Academic Press, New York (1973)
61. Manski, C.F.: Maximum score estimation of the stochastic utility model of choice. J. Econ. **3**(3), 205–228 (1975)
62. Jackson, M., Watts, A.: The existence of pairwise stable networks. Seoul J. Econo. **14**(3), 299–321 (2001)

Networks and Context: Algorithmic Challenges for Context-Aware Social Network Research

Mirco Schoenfeld and Juergen Pfeffer

Abstract Social interaction is mediated by computer processes at an ever-increasing rate not least because more and more people have smartphones as their everyday and habitual companions. This enables collection of a vast amount of data containing an unprecedented richness of metadata of interaction and communication. Such context information contains valuable insights for social network research and allows for qualitative grading of network structure and consecutive structural analysis. Due to the complex, heterogeneous, dynamic, and uncertain nature of such information it is yet to be considered for network analysis tasks in its entirety. In this paper, we emphasize how network analysis benefits from considering context information and identify the key challenges that have to be tackled. From an algorithmic perspective, such challenges appear on all steps of a network analysis workflow: Dynamics and uncertainty of information affects modeling networks, calculation of general metrics, calculation of centrality rankings, graph clustering, and visualization. Ultimately, novel algorithms have to be designed to combine context data and structural information to enable future context-aware network research.

Keywords Context awareness · Social networks · Network research

1 Context-Aware Social Network Research

Social interaction is happening in computer-assisted environments more and more—humans cultivate their friendships, organizations collaborate professionally, and more and more people carry around small computers to do phone calls. Regardless of what entities are communicating, online social interaction is mediated by computer processes. Such systems offer implicit access to social networks and the

M. Schoenfeld (✉) · J. Pfeffer
Bavarian School of Public Policy, Technical University in Munich, Munich, Germany
e-mail: mirco.schoenfeld@tum.de; juergen.pfeffer@tum.de

© Springer Nature Switzerland AG 2020
G. Ragozini, M. P. Vitale (eds.), *Challenges in Social Network Research*,
Lecture Notes in Social Networks, https://doi.org/10.1007/978-3-030-31463-7_8

content that is shared in such networks. At the same time, such systems increasingly gain access to metadata of communication that describes context of entities, their actions, and interactions. Such context information is yet to be considered for network analysis tasks.

We consider context information to be complex, heterogeneous, dynamic, and incomplete information. In that, it is more than a fixed number of static attributes annotating edges or vertices. It is instead highly varying information with uncertainty.

With context-aware social network research we aim at analyzing co-evolution of network structure and behavior of entities of networks. Ultimately, we want to investigate mutual influences between individuals and their surroundings in a network.

Network analysis focuses mostly on rankings among nodes and subgroup membership, that is, quantitative analysis of network structure. For such tasks there is a considerable number of highly optimized tools and algorithms readily available [46]. Nevertheless, all of these approaches require networks being modeled by operationalized simplification such that metaphors for nodes and edges are unambiguous. That leads to context information being aggregated or even omitted. However, it is only context information that allows for qualitative grading of modeled structure and consecutive structural analysis.

In interpersonal networks, for example, the mere existence of an edge between two humans contains very little information about the quality or intensity of a relationship. Assessing that is only possible if information about time, location, type, degree of kindness, or even warmth is considered—all of which are descriptions of the context in which interaction takes place [42].

But, current analysis methods lack a possibility to thoroughly integrate such information. A well-known way for network researchers is to involve some external information besides pure structural information is using edge weights. While it might be suitable to consider edge weights as degree of kindness of interaction (or other types of data that allow for a quantitative mapping), the metaphor certainly is inappropriate for encoding location, motivations, or intentions behind interactions, let alone combinations of such data.

We claim that a novel way of thinking is required for network researchers which we want to introduce and motivate with this paper: *context-aware social network research*. In this envisioned area of research, algorithms and methods will be developed that enable network researchers to augment quantitative network analysis by means of complex and diverse data. We encourage network researchers to collect context information alongside the usual information that will be translated into network structure. Context-aware social network research is about using this metadata to augment network analysis like it is shown in Fig. 1. Ultimately, context information enables a better quantification of valuable qualitative insights opening a mixed perspective of both qualitative and quantitative research on social network analysis [18].

To provide a clear understanding, we start by giving a definition of context for context-aware social network research in Sect. 2. Based on this, we illustrate

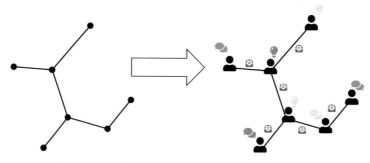

Structure-only network analysis   Context-aware network analysis

Fig. 1 Context awareness augments structural network analysis with additional information on edges (like location) and nodes (like content from Social Media, demographic information, and more) to enable novel qualitative insights

our perception of integrating a comprehensive contextual perspective by giving concrete examples of networks with annotated nodes and edges in Sect. 3. Then, in Sect. 4, we emphasize how network analysis will benefit from considering this augmenting information on all steps of a network analysis workflow. In Sect. 5 we specify the key algorithmic challenges that currently hinder an integration of an all-encompassing view on context into algorithms for network analysis. Section 6 concludes this paper.

2 Definition of Context

Mentioning context and context-aware algorithms usually addresses a specific field of research. It was in 1994, when the term *context-aware computing* was coined which introduced a novel way for algorithms to adapt to the real world. Originally, the term describes a computer's ability to adapt to its location of use and to objects in close proximity [39]. Since then, a lot of effort has been put into formulating a proper definition of what context actually is. One of the available definitions that is widely accepted for its simplicity and its general scope was proposed by Dey:

> Context is any information that can be used to characterise the situation of an entity. An entity is a person, place, or object that is considered relevant to the interaction between a user and an application, including the user and applications themselves. [11]

With this definition a lot of previous work, many of which lacked a clear distinction between "situation" and "environment," found a common ground.

Nevertheless, this definition leaves several central elements undefined and thus incorporates common sense about what actually is the concrete meaning of terms such as information and situation. This can be explained from the underlying technical, application-centric perspective.

For the early years of context-aware computing with few and simple sensors and limited computing capabilities this was a pragmatic perspective. But, since 1994, variety and reliability of sensors as well as computing capabilities have increased significantly. Modern approaches on context take into account the fact that relevant context information can be more than mere sensor readings: context should be divided into context atoms and higher-level context where context atoms represent simple sensor readings, while higher-level context should be seen as rich knowledge deduced from a continuous stream of heterogeneous information contrasting sensor readings with prior knowledge as well as information from external sources [47].

For the scope of context-aware social network research however, we claim that the above-given definition and clarifications of context still fall short in meeting unique requirements of this field of research. In order to support future developments, we want to extend and concretize the definition of context as follows:

Definition 1 Context is any information that can be used to characterise an entity. An entity can be a node, an edge, or a network in general. Information characterising an entity is all information that is not used to model the structure of the network.

This definition takes into account the discriminative process of deciding which part of collected data is considered useful for modeling a network and which part is used as additional information. Hennig et al. call this additional information "actor-attributes" and "tie-attributes," whereas Scott generalizes such information as "attribute data" [16, 40]. Both works use their respective terminology to discriminate attribute data from "relational data" which contains structural information.

However, existing perspectives on context information which use the terminology of attribute data lack the unrestrictedness of what context can be. Instead, the above-mentioned terminology associates itself to the mental construct of tables and spreadsheets implying the necessity of compressing additional information such that it can be incorporated as an edge weight or a node type because this is basically what well-known network analysis algorithms allow [46].

We want to emphasize that our above-given definition does not restrict the *type* of information that describes an entities' context. Instead, we consider all complex, heterogeneous, dynamic, incomplete, and time-dependent information to be a valid part of a thorough context description.

With this in mind, we show that novel methods are necessary that support the involvement of complex and heterogeneous information into analysis tasks. Thus, the following section illustrates some scenarios in which context will be beneficial for common analysis tasks.

3   Contextualized Networks

Since this work identifies algorithmic challenges for integrating context into network analysis, this section illustrates some of the various manifestations of relevant context information for social network research by giving some examples of *contextualized networks* in which context information is available to nodes, edges, or both in the network.

3.1 Context in Interpersonal Networks

Friendships or family relations are often formed because individuals share common motives, ideologies, or personality traits—the *homophily* principle [30]. Enriching the resulting network structure with a description of individuals as comprehensive as possible can lead to insights into the foundations of relations. Such additional description then serves as context information illustrating the nodes of the network supporting insights into the quality of connecting relations.

Social network research also often considers an Online Social Network (OSN) as an object of investigations. Naturally, the nodes of the corresponding network are profiles registered in the OSN. It can be of interest to let the edges represent interactions between these profiles. For example, a directed edge between profile A and profile B is only present if A commented on a piece of content that was produced by B earlier. A similar example is a network of eMail exchange in which nodes represent involved eMail accounts and edges represent eMails. In both scenarios, insights into the nature of the interaction can be gained by considering the content of the interaction like texts, images, or even videos—all of which can serve as context information illustrating the edges.

3.2 Context in Document Networks

A corpus of written documents often contains valuable network information. An obvious example is the literature section in scientific papers describing an essential part of a citation network [21]. In such a network, nodes represent papers which are connected if a paper appears as a citation in another paper.

A context information that is often disregarded in the analysis of citation networks is the content of the papers. The content contains a certain set of ideas, a certain set of topics, and even a certain style of argumentation all of which serve as illustrative context information about the nodes. As such, this information may help in describing the various facets of a field of research and its diverse connection points to other fields, for example.

A corpus of documents may contain other kinds of networks as well. Considering a corpus of newspaper articles, co-mentioning of politicians in the same article can easily be translated into a network of politicians.

As an example, the relational information can be augmented with the textual surrounding in which both names appeared. Contextualizing the edges with this information will help differentiating the topics or the sentiment under which the actors in question appear for public perception.

Going even further, textual surroundings also implicitly induce associations to related meanings and semantics. Such information is inherently complex itself.

3.3 The Outside World as Context

Social sciences will increasingly devote itself to datasets compiled by pervasive and ubiquitous applications. Such applications are most common on mobile phones since these are becoming permanent companions. But such applications will be found in Internet-of-Things devices in the near future just as well.

These devices enable the collection of context information at the very moment an interaction with an application is taking place. Consider an application for mobile social networking—this could be a counterpart of a social networking website available as a native application for a mobile device. From an application provider's perspective, the underlying social network can be enriched with valuable context information. On the one hand, nodes can be annotated with locations visited frequently or recently. On the other hand, edges can be annotated by time and location at which an interaction is taking place. Of course, modern mobile devices offer a wide range of additional sensors that could be taken into account depending on the use case of an application and an analysis task. As such, sensors measuring physical activity or types of movement come to mind. Also, a mobile phone's media player can supply context information in terms of the kind of music that accompanies certain locations, times of day, or interactions. However, time and location alone already allow for detecting certain geographic or habitual preferences in interacting with peers.

To summarize, all of the scenarios described above show how mere structural network information can be enriched by additional context information which will support important analysis tasks beneficially. The following section will make precise distinction on how the structural analysis can benefit from such additional information.

4 Analytical Potential and Related Work

This section clarifies how social network research will benefit from incorporating context information into network analysis methods. Since different methodological aspects of social network research are affected, this section is structured following a typical network research procedure. This means to start with creating a network from the data in question, then analyze this network by means of centrality measurements, community detection, or other network-analytical approaches, and finally, visualize a network in order to communicate findings. At every step of the procedure, related work is mentioned.

4.1 Network Creation

The step of modeling a network usually requires to identify certain parts of collected data containing a complex structure that are worth being represented as a network. The rest of the data is then discarded for the following analysis of structure. But, from an analytical perspective, it could be beneficial to not discard the remaining information but to consider it as context information and also model a network from it.

A natural way of modeling a network comprising multiple contextual facets is based on *multiplex* or *multilayer* networks. These are networks comprising of sets of overlapping networks or layers [5, 26]. Each layer corresponds to a category of context such that nodes are connected differently on each layer. Such a network has just as many layers as distinct types of context information.

An alternative approach is to identify concepts or abstractions of context and use these as nodes of a *two-mode-* or *affiliation*-network. Such networks combine nodes of two (or more) different types in one network representation [15, 27].

Both approaches face considerable limitations: The number of context dimensions renders multilayer networks resource-consuming and inefficient. Further, since edges have to be instantiated based on thresholding some similarity metric, resulting network layers either lack resolution or discriminative power. The same holds for modeling nodes as abstractions of context for an affiliation network.

To overcome these obstacles, one would model an annotated graph [25]. In such a network model, additional information is kept for both nodes and edges. This information is then readily available during subsequent analysis steps and the following sections will present related analysis methods. In general, however, analysis of annotated networks currently deals with pre-given categories of annotations or feature vectors of fixed size which would be insufficient to model our understanding of context.

An additional level of complexity is induced by the dynamic nature of modeled systems. In social networks, for example, friendship relations change as well as situations of individuals. With respect to a network model this requires a model to support structural dynamics as well as changes in annotations. It is only recently that research considers such highly variational settings [28, 34].

4.2 Network-Level Metrics

Having context information associated with elements of a network enables a deeper evaluation of common network metrics. Consider, for example, **reciprocity** which is a measure of the likelihood of nodes in a directed network being mutually linked. Such a metric would greatly benefit if mutuality of edges could be evaluated in terms of contexts: Are edges reciprocal in similar contexts? Does cross-contextual reciprocity have a different effect compared to reciprocity within one context?

A similar evaluation would be possible for **transitivity**: Does transitivity span contexts or are certain relations transitive within one context? What does it mean for a certain contextual setting that transitivity is measurable across settings? Are the contextual settings so diverse that it does have an effect? This has been evaluated for multiplex networks and scenarios with a limited number of context dimensions such that every dimension can be encoded into a network [1].

Another metric would be the evaluation of **assortativity**. For certain networks it could be illuminating to examine if contextual properties of nodes govern the process of attaching to other nodes rather than a node's degree only. Related work in the field is often concerned with correlating structural information and attributes of nodes [33, 35, 37]. Supported complexity and dynamics of attributes is limited, though.

4.3   Node-Level Metrics

A related family of network analysis algorithms tackles the problem of **centrality calculation**. Here, we see two possible research directions. On the one hand, one can evaluate nodes' centralities for separate contexts in order to identify variations between contextual settings. This would clarify if prominent nodes are prominent across contexts or if there is a skewness in centrality values towards a certain context. This has been investigated for multilayer networks to some extent [43, 44]. However, computational complexity renders these approaches unsufficient for a larger number of context dimensions.

On the other hand, one can correlate centrality measurements with the influence of contexts. For a longitudinal network study evaluating the development of centrality values of some nodes, for example, it could be of interest to observe the influence of respective contexts. This would help to identify those contextual settings that cause noticeable changes in centrality values.

4.4   Community Detection

The problem of graph clustering or **community detection** benefits greatly from involving context information. In the most simple setting of a single static network nodes are pooled into communities because of variation in density or modularity for certain parts of a network. This has been extensively extended to scenarios in which additional relational information is available and multilayer networks are modeled [10, 13, 17, 23].

In recent years, considering more than structural information has led to detecting **overlapping communities** in annotated graphs [22]. Figure 2 depicts overlapping communities as a mixture of structural and additional information. The left part of the figure shows the structural notion of community that is based on variations in

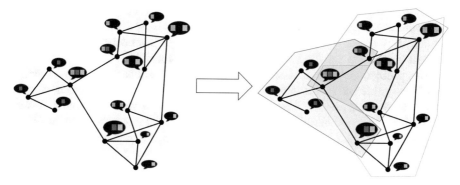

Fig. 2 Community detection on a contextualized network based on node annotations. The left part shows the procedure based on structural information alone. The right part shows how involvement of node annotations could change community notation

density between inside and outside of communities and the right part of the figure shows how communities can overlap containing nodes of similar annotations as well.

In early works, annotations were considered fixed categories or feature vectors of fixed size and network structure remained constant [7, 8, 14, 31]. For such overlapping communities, the interplay between clusters in graph structure and clusters in annotations has been of interest [9]. Later, dynamic graphs were considered in which structure changed but attributes remained static [6, 38]. See [22] for a recent overview over community detection methods in annotated networks.

It is only recently that both dynamics in network structure and dynamics in annotations are investigated in terms of overlapping communities [3, 20, 32]. However, assumptions about annotations do not fully cover our definition of context information as, for example, uncertainty of such information is mostly disregarded.

4.5 Visualizing Networks

Since network analysis is often about communicating a complex finding by means of an expressive visualization, it is important to include context information into the final visual presentation of the message as well. In fact, this is a well-known practice for networks in which a single context information with an appropriate measurement for similarity is available; networks of airports and connecting flights are often drawn on maps, for example—airports have a certain geographic location associated which consequently steers the placement of the corresponding nodes, while the underlying maps imply the translation of geographic distance into a distance on the given canvas.

Of course, layouting a network is not only about placing nodes on a canvas [36], especially, if edges are annotated with certain context information as well. In this case, it is vital to visually encode context similarities for edges meaning edges with similar context annotations must be distinguishable from those annotated with strongly differing contextual settings. Or, in other words, **multiplexity of edges** in context-augmented networks needs to be visually encoded [12].

This problem gets more complex for increasing dynamics in network structure which has gained attention in recent years [2]. For additional dynamics in context attributes, however, proper visualization techniques have yet to be found. Visual discriminability particularly ensures that valuable context information are available for the last part of the network analysis procedure: the interpretation of results.

4.6   Interpretation of Network Analysis

The last part of a typical network analysis procedure is interpretation of results. Considering a network that has been created and analyzed with the help of one or more context-aware methods outlined above we see a substantial benefit for the interpretation of network-analytical findings: involving context information on several steps of the analysis procedure allows better quantitative description of qualitative information and unveils qualitative aspects of structural relations [18].

As we see it, context-aware network analysis methods can enable researchers to acquire a better sense of how edges emerge and develop, of how nodes interact with each other, and of how relevance develops. Ultimately, such methods allow to investigate the influence of network surroundings on individual nodes and the influence of individual nodes on surroundings in the network.

5   Algorithmic Challenges

To tap the full potential of context information for network analysis such data has to be made available in a way that algorithms can process it effectively and efficiently. One obstacle for effectively processing context information is the varying nature of such data in terms of data sources, data types, and meaning. The efficient processing on the other hand could be hindered by the size of such data. Depending on the use-case context information could of course be numeric or textual data. Especially for research regarding Online Social Networks (OSNs), such multimedia data is a commonly used way of expressing oneself.

Thus, for our envisioned methods to be successful, we identify the following key challenges:

- Heterogeneity of data types and data sources
- Organization of combined structural and context data
- Adjustment of influence of annotations

- Linking to raw data
- Missing data

The remainder of this section details the above-listed challenges.

5.1 Heterogeneity of Data Types and Data Sources

Dealing with context information is not a trivial task. By definition, there is no restriction on allowed data types or data sources. As Sect. 2 pointed out context data can be raw sensor readings—context atoms—as well as higher-level context which in turn can be composed of context atoms or some other kind of inference mechanism performed on the latter. This introduces uncertainty and missing information. Thus, the major challenge is determining (dis-)similarity of context information.

Most of the envisioned approaches mentioned in Sect. 4 rely on comparison of context information at some point: Be it to determine cross-contextual transitivity, to evaluate clustering of nodes, or to infer layout information for node placement or edge distinction. As long as context information is composed of context atoms alone, comparison of such information can be evaluated using a distance metric appropriate for the respective sensor. When comparing location descriptions as reported by a GPS sensor, for example, geographic positions are usually translated into WGS84 coordinates and compared using Karney's method [24]. But, for higher-level descriptions of context an appropriate distance function is often not that obvious to select. Further, having a single distance function for every type of context data induces the *curse of dimensionality*.

One generalization of context data that hides heterogeneity of data types and data sources to subsequent processing steps is using Bloom filters [4]. This technique has been used to encode context information to establish a context-centric social network [47]. The major advantage of this coding scheme is that it requires only one distance function for an arbitrary number of data modalities. All context-based calculations are evaluated on the Bloom filter which summarizes the actual context information and thereby decouples the representation of information from the underlying data types and sizes of data objects, completely.

The downside of this approach is that it is a probabilistic encoding that is prone to loss of information. Depending on the configuration of Bloom filters representation of information can be ambiguous to a certain extent.

5.2 Organization of Combined Structural and Context Data

Enriching networks with context data requires sophisticated indexing techniques with respect to storage and organization of data. Most likely, complex and hetero-geneous context data will be kept separate from structural information. This allows efficient processing of network structure and prevents arbitrary large context data from being copied to main memory and back. Nevertheless, all of the envisioned methods and algorithms from Sect. 4 need access to context information at some point—often even random access is required to reach acceptable runtimes. This is the case for layouting a network such that nodes with similar context descriptions should be placed closer to each other, for example. Any two nodes with similar context descriptions are not necessarily linked to each other which could be used to predict access to their respective context information to some extent.

A feasible way to solve this may be to use Bloom filters as representations of context information. They provide a compact summary of arbitrary sets with a constant and adjustable size which may fit into main memory together with structural information [47].

Appropriate organization and indexing of data will be a concern when dealing with dynamic data. In OSNs or mobile scenarios, for example, certain context information tend to change a lot. In this case, the use of Bloom filters has to be deliberated since they do not support deletion of information. Instead, it might be beneficial to incorporate an alternative representation or addressing scheme for context information which makes use of Locality Sensitive Hashing (LSH) or similar hashing procedures together with an Approximate Nearest Neighbor (ANN) search [29, 45].

5.3   Adjustment of Influence of Annotations

Since context information is only meant to augment network research it is important to adjust the extent to which context information supervises structural analysis. For example, regarding the task of community detection in networks the degree of augmentation would reach from considering only structural information up to observing only context information if the degree of augmentation could be specified in a range from 0 to 100%. When context information contributes close to zero percent community detection would function in the way of comparing variance of structural modularity. When, on the other hand, context information contributes to the maximum extent of 100% the problem of community detection transforms to a problem of clustering nodes based on their contextual properties alone.

To emphasize: the degree of augmentation has to be adjustable for network researchers who have to decide based on concrete use-cases.

5.4   Linking to Raw Data

Augmentation of structural information with context data is meant to enable qualitative evaluation and interpretation of network analysis results. For this qualitative perspective to be profound and meaningful providing access to actual context data that is related to algorithmic output is mandatory. This concerns basic metrics like assortativity as well as more complex analysis methods like centrality calculation. Only if actual characteristics of context are accessible a researcher is able to comprehend mechanisms of preferential attachment or processes of gaining importance.

This necessity imposes strong requirements on data management of structural data together with context information as it was described in Sect. 5.2.

Further, special care has to be taken if analysis methods rely on probabilistic context coding schemes. In fact, this requires transparency and confirmability of causal effects of certain parts of context summaries: If a probabilistic summary consists of Bit arrays it may be complicated to identify those context properties that had an actual effect on calculation of assortativity.

5.5 Missing Data

Since our definition of context allows for information to be dynamic and incomplete, envisioned methods have to be able to handle missing information [41]. A feasible approach would be to use context information to infer missing links, missing nodes, or missing context information. Comparable inference mechanisms have been proposed in [19, 37].

Another approach would be to integrate an understanding of uncertainty into the general network model such that all steps of the analysis are affected. If, for example, context is modeled using probabilistic coding schemes uncertainty is inherently present. This would pose interesting questions for calculating rankings, clusters, and visualizations.

6 Conclusion

With this paper we outline our vision of context-aware social network research which is about enabling methods of network analysis to involve context information as annotations for nodes and/or edges. These novel algorithms can unveil qualitative insights into complex data and induce a novel way of thinking about network research.

Context-aware social network research differs from existing work on annotated networks or multilayer networks in the definition of context information. Hence, we outline our understanding of context being complex, heterogeneous, dynamic, and incomplete information. In that, it is more than a fixed number of static attributes annotating edges or vertices. It is instead highly varying information with uncertainty.

Based on our definition of context, we detail how considering such information would benefit individual steps along a typical network analysis procedure. From its potential we derive specific challenges for the development of novel envisioned analysis methods which arise from the heterogeneous nature of context data as well as from the expected amounts of context data.

Ultimately, the envisioned combination of structural and contextual data can enable a deeper understanding of how edges emerge and form, of how nodes interact with each other, and of where structural importance originates.

References

1. Battiston, F., Nicosia, V., Latora, V.: Structural measures for multiplex networks. Phys. Rev. E **89**, 032804 (2014). https://doi.org/10.1103/PhysRevE.89.032804
2. Beck, F., Burch, M., Diehl, S., Weiskopf, D.: A taxonomy and survey of dynamic graph visualization. Comput. Graphics Forum **36**(1), 133–159 (2017). https://doi.org/10.1111/cgf.12791
3. Bello, G.A., Harenberg, S., Agrawal, A., Samatova, N.F.: Community detection in dynamic attributed graphs. In: Li, J., Li, X., Wang, S., Li, J., Sheng, Q.Z. (eds.) Advanced Data Mining and Applications, pp. 329–344. Springer, Cham (2016)
4. Bloom, B.H.: Space/time trade-offs in hash coding with allowable errors. Commun. ACM **13**(7), 422–426 (1970). https://doi.org/10.1145/362686.362692
5. Boccaletti, S., Bianconi, G., Criado, R., del Genio, C., Gómez-Gardenes, J., Romance, M., Sendina-Nadal, I., Wang, Z., Zanin, M.: The structure and dynamics of multilayer networks. Phys. Rep. **544**(1), 1–122 (2014). https://doi.org/10.1016/j.physrep.2014.07.001
6. Bojchevski, A., Günnemann, S.: Bayesian robust attributed graph clustering: joint learning of partial anomalies and group structure. In: Thirty-Second AAAI Conference on Artificial Intelligence (2018)
7. Bothorel, C., Cruz, J.D., Magnani, M., Magnani, B.: Clustering attributed graphs: models, measures and methods. Netw. Sci. **3**(3), 408–444 (2015). https://doi.org/10.1017/nws.2015.9
8. Bu, Z., Gao, G., Li, H.J., Cao, J.: CAMAS: a cluster-aware multiagent system for attributed graph clustering. Inform. Fusion **37**, 10–21 (2017). https://doi.org/10.1016/j.inffus.2017.01.002
9. Cheng, H., Zhou, Y., Yu, J.X.: Clustering large attributed graphs: a balance between structural and attribute similarities. ACM Trans. Knowl. Discov. Data **5**(2), 12:1–12:33 (2011). https://doi.org/10.1145/1921632.1921638
10. De Bacco, C., Power, E.A., Larremore, D.B., Moore, C.: Community detection, link prediction, and layer interdependence in multilayer networks. Phys. Rev. E **95**, 042317 (2017). https://doi.org/10.1103/PhysRevE.95.042317
11. Dey, A.K.: Understanding and using context. Pers. Ubiquit. Comput. **5**(1), 4–7 (2001). https://doi.org/10.1007/s007790170019
12. Domenico, M.D., Porter, M.A., Arenas, A.: MuxViz: a tool for multilayer analysis and visualization of networks. J. Complex Netw. **3**(2), 159–176 (2015). https://doi.org/10.1093/comnet/cnu038
13. Du, Y., Gao, C., Chen, X., Hu, Y., Sadiq, R., Deng, Y.: A new closeness centrality measure via effective distance in complex networks. Chaos Interdiscip. J. Nonlinear Sci. **25**(3), 033112 (2015). https://doi.org/10.1063/1.4916215
14. Günnemann, S., Färber, I., Raubach, S., Seidl, T.: Spectral subspace clustering for graphs with feature vectors. In: 2013 IEEE 13th International Conference on Data Mining, pp. 231–240 (2013). https://doi.org/10.1109/ICDM.2013.110

15. Gong, N.Z., Xu, W., Huang, L., Mittal, P., Stefanov, E., Sekar, V., Song, D.: Evolution of social-attribute networks: measurements, modeling, and implications using Google+. In: Proceedings of the 2012 Internet Measurement Conference (IMC '12), pp. 131–144. ACM, New York (2012). https://doi.org/10.1145/2398776.2398792
16. Hennig, M., Brandes, U., Pfeffer, J., Mergel, I.: Studying Social Networks: A Guide to Empirical Research. Campus Verlag, Frankfurt (2012)
17. Hmimida, M., Kanawati, R.: Community detection in multiplex networks: a seed-centric approach. Networks Heterog. Media 10(1), 71–85 (2015). https://doi.org/10.3934/nhm.2015.10.71
18. Hollstein, B.: Qualitative approaches. In: Scott, J., Carrington, P.J. (eds.) The SAGE Handbook of Social Network Analysis, pp. 404–416. SAGE Publications, London (2014)
19. Hric, D., Peixoto, T.P., Fortunato, S.: Network structure, metadata, and the prediction of missing nodes and annotations. Phys. Rev. X 6, 031038 (2016). https://doi.org/10.1103/PhysRevX.6.031038
20. Hulovatyy, Y., Milenkovic, T.: Scout: simultaneous time segmentation and community detection in dynamic networks. Sci. Rep. 6, 37557 (2016)
21. Hummon, N.P., Doreian, P., Freeman, L.C.: Analyzing the structure of the centrality-productivity literature created between 1948 and 1979. Knowledge 11(4), 459–480 (1990). https://doi.org/10.1177/107554709001100405
22. Javed, M.A., Younis, M.S., Latif, S., Qadir, J., Baig, A.: Community detection in networks: a multidisciplinary review. J. Netw. Comput. Appl. 108, 87–111 (2018). https://doi.org/10.1016/j.jnca.2018.02.011
23. Jeub, L.G.S., Mahoney, M.W., Mucha, P.J., Porter, M.A.: A local perspective on community structure in multilayer networks. Netw. Sci. 5(2), 144–163 (2017). https://doi.org/10.1017/nws.2016.22
24. Karney, C.F.F.: Algorithms for geodesics. J. Geodesy 87(1), 43–55 (2013). https://doi.org/10.1007/s00190-012-0578-z
25. Kim, M., Leskovec, J.: Multiplicative attribute graph model of real-world networks. In: Kumar, R., Sivakumar, D. (eds.) Algorithms and Models for the Web-Graph, pp. 62–73. Springer, Berlin (2010)
26. Kivelä, M., Arenas, A., Barthelemy, M., Gleeson, J.P., Moreno, Y., Porter, M.A.: Multilayer networks. J. Complex Netw. 2(3), 203–271 (2014). https://doi.org/10.1093/comnet/cnu016
27. Leifeld, P.: Discourse network analysis: policy debates as dynamic networks. In: The Oxford Handbook of Political Networks. Oxford University Press, Oxford (2017). https://doi.org/10.1093/oxfordhb/9780190228217.013.25
28. Li, J., Dani, H., Hu, X., Tang, J., Chang, Y., Liu, H.: Attributed network embedding for learning in a dynamic environment. In: Proceedings of the 2017 ACM on Conference on Information and Knowledge Management (CIKM '17), pp. 387–396. ACM, New York (2017). https://doi.org/10.1145/3132847.3132919
29. Masci, J., Bronstein, M.M., Bronstein, A.M., Schmidhuber, J.: Multimodal similarity-preserving hashing. IEEE Trans. Pattern Anal. Mach. Intell. 36(4), 824–830 (2014). https://doi.org/10.1109/TPAMI.2013.225
30. McPherson, M., Smith-Lovin, L., Cook, J.M.: Birds of a feather: homophily in social networks. Ann. Rev. Sociol. 27(1), 415–444 (2001). https://doi.org/10.1146/annurev.soc.27.1.415
31. Mei, Q., Cai, D., Zhang, D., Zhai, C.: Topic modeling with network regularization. In: Proceedings of the 17th International Conference on World Wide Web (WWW '08), pp. 101–110. ACM, New York (2008). https://doi.org/10.1145/1367497.1367512
32. Meng, L., Hulovatyy, Y., Striegel, A., Milenković, T.: On the interplay between individuals' evolving interaction patterns and traits in dynamic multiplex social networks. IEEE Trans. Netw. Sci. Eng. 3(1), 32–43 (2016). https://doi.org/10.1109/TNSE.2016.2523798
33. Newman, M.E.J.: Mixing patterns in networks. Phys. Rev. E 67, 026126 (2003). https://doi.org/10.1103/PhysRevE.67.026126

34. Nguyen, G.H., Lee, J.B., Rossi, R.A., Ahmed, N.K., Koh, E., Kim, S.: Continuous-time dynamic network embeddings. In: Companion Proceedings of the Web Conference 2018 (WWW '18), pp. 969–976. International World Wide Web Conferences Steering Committee, Republic and Canton of Geneva (2018). https://doi.org/10.1145/3184558.3191526
35. Pelechrinis, K., Wei, D.: VA-index: Quantifying assortativity patterns in networks with multidimensional nodal attributes. PLoS One **11**(1), 1–13 (2016). https://doi.org/10.1371/journal.pone.0146188
36. Pfeffer, J.: Visualization of political networks. In: Victor, J.N., Montgomery, A.H., Lubell, M. (eds.) The Oxford Handbook of Political Networks. Oxford University Press, Oxford (2017). https://doi.org/10.1093/oxfordhb/9780190228217.013.13
37. Rabbany, R., Eswaran, D., Dubrawski, A.W., Faloutsos, C.: Beyond assortativity: proclivity index for attributed networks (prone). In: Kim, J., Shim, K., Cao, L., Lee, J.G., Lin, X., Moon, Y.S. (eds.) Advances in Knowledge Discovery and Data Mining, pp. 225–237. Springer, Cham (2017)
38. Ranshous, S., Shen, S., Koutra, D., Harenberg, S., Faloutsos, C., Samatova, N.F.: Anomaly detection in dynamic networks: a survey. Wiley Interdiscip. Rev. Comput. Stat. **7**(3), 223–247 (2015). https://doi.org/10.1002/wics.1347
39. Schilit, W.N.: A system architecture for context-aware mobile computing. Ph.D. Thesis, Columbia University (1995)
40. Scott, J.: Social Network Analysis. Sage, Thousand Oaks (2017)
41. Sharma, R., Magnani, M., Montesi, D.: Investigating the types and effects of missing data in multilayer networks. In: 2015 IEEE/ACM International Conference on Advances in Social Networks Analysis and Mining (ASONAM), pp. 392–399 (2015). https://doi.org/10.1145/2808797.2808889
42. Snijders, T., Steglich, C., Schweinberger, M.: chap. Modeling the Coevolution of Networks and Behavior, pp. 41–71. Lawrence Erlbaum Associates Publishers, Mahwah (2007)
43. Solá, L., Romance, M., Criado, R., Flores, J., García del Amo, A., Boccaletti, S.: Eigenvector centrality of nodes in multiplex networks. Chaos Interdiscip. J. Nonlinear Sci. **23**(3), 033131 (2013). https://doi.org/10.1063/1.4818544
44. Solé-Ribalta, A., De Domenico, M., Gómez, S., Arenas, A.: Centrality rankings in multiplex networks. In: Proceedings of the 2014 ACM Conference on Web Science (WebSci '14), pp. 149–155. ACM, New York (2014). https://doi.org/10.1145/2615569.2615687
45. Wang, J., Liu, W., Kumar, S., Chang, S.F.: Learning to hash for indexing big data—a survey. Proc. IEEE **104**(1), 34–57 (2016). https://doi.org/10.1109/JPROC.2015.2487976
46. Wasserman, S., Faust, K.: Social Network Analysis: Methods and Applications. Cambridge University Press, Cambridge (1994)
47. Werner, M., Dorfmeister, F., Schönfeld, M.: Ambience: context-centric online social network. In: 12th Workshop on Positioning, Navigation and Communications (WPNC) (2015)

Part II
Applications

Unraveling Innovation Networks in Conservation Agriculture Using Social Network Analysis

Juan Manuel Aguirre-López, Julio Díaz-José, Petra Chaloupková, and Francisco Guevara-Hernández

Abstract During the last decades, agriculture has focused on developing more sustainable forms of land use while promoting food productivity. To face these challenges, conservation agriculture (CA), which is based on minimal soil disturbance, mulching of crop residues, and crop rotation, has been promoted as an ecosystem approach to sustainable agriculture. Using social network analysis (SNA) methods, we analyzed the adoption patterns of CA practices among 222 maize smallholder farmers in the Mexican state of Chiapas. Our findings suggest that, in the process of adopting CA practices, farmers make interrelated decisions based on network and individual attributes, rather than accepting top-down transfers, as usually promoted by institutions. The role of extension agents and other farmers is crucial for innovation dissemination, given that farmers learn different practices from different sources. This points to the need to strengthen participatory methods and promote sustainability in agriculture, rather than apply the usual hierarchical mechanisms in the innovation dissemination process.

Keywords Social network analysis · Innovation adoption · Conservation agriculture · Agricultural sustainability

J. M. Aguirre-López
Universidad Autónoma Chapingo, Chapingo, Mexico

J. Díaz-José (✉)
Tecnológico Nacional de México-Campus Zongolica, Veracruz, Mexico

Facultad de Ciencias Biológicas y Agropecuarias, Universidad Veracruzana, Veracruz, Mexico

P. Chaloupková
Czech University of Life Sciences, Prague, Czech Republic

F. Guevara-Hernández
Universidad Autónoma de Chiapas, Villaflores, Chiapas, México

© Springer Nature Switzerland AG 2020
G. Ragozini, M. P. Vitale (eds.), *Challenges in Social Network Research*,
Lecture Notes in Social Networks, https://doi.org/10.1007/978-3-030-31463-7_9

1   Introduction

In Mexico, maize (*Zea mays* L.) is grown on 7.4 million ha, with an average yield of 3.3 kg ha^{-1}, which in 2016 resulted in the production of 28.25 million tons [1]. The estimated per capita maize consumption in the country is nearly 267 g/person/day [2], making this grain the main source of protein and calories for the human population in poor rural areas [3]. Thus the importance of maize in Mexico goes far beyond its economic benefits, which may seem to break with economic rationality, given the social significance of maize and the opportunity that it offers for the livelihoods of most of the rural population [4, 5]. Given the importance of this crop, institutions are actively promoting more sustainable agriculture for producing maize in order to maintain the advantages that ecological agricultural practices provide. In recent years, agricultural system sustainability has focused on the need to develop new technologies and practices that address extensive soil degradation, are effective and accessible to farmers, and improve agricultural productivity [6–8]. To face these challenges, conservation agriculture (CA)—"a farming system that promotes maintenance of a permanent soil cover, minimum soil disturbance (i.e. no tillage), and diversification of plant species" (http://www.fao.org/conservation-agriculture/)—is widely promoted as an option of sustainable agriculture. However, challenges remain for the adoption of CA principles, especially in the case of poor farmers [9] who require complementary practices to make CA a more functional system for guaranteeing success when applying these practices [10].

In Mexico, CA practices have been promoted by research institutions such as the International Maize and Wheat Improvement Center (CIMMYT), the National Forestry, Agricultural and Livestock Research Institute (INIFAP), and other government initiatives such as the Trust Funds for Rural Development (FIRA). Using different extension strategies, those institutions are actively promoting the adoption of CA principles by smallholder farmers in regions such as Chiapas. However, during recent decades, new actors and institutions have entered agricultural production and markets, which suggests the need to change less linear agricultural extension approaches[1] for more sophisticated methods, in order to understand and exploit new opportunities [11]. In this way, new approaches such as agricultural innovation systems (AIS) have become important for understanding challenges and identifying opportunities to improve the innovation capacity of agricultural systems [12], in which diverse actors contribute to the innovation process. Thus, the AIS approach can contribute to addressing discrepancies between policy changes, actors and their relationships, and productivity issues [13]. CA adoption process is most efficient when a local AIS emerges and begins to acquire a self-sustaining dynamic as a result of the interaction between different types of actors [9]. Several studies have highlighted the importance of analyzing social networks, because to

---

[1]Linear extension models increase agricultural production through technology transfer and adoption, if research addressed farmers' problems and using a "supply" orientation for delivering new technologies. However, this approach has been criticized for being inefficient and poorly targeted.

understand agricultural adoption rates, it is necessary not only to have information about individual attributes but also about social contacts [14–17].

As stated above, agricultural innovation studies have brought out the importance of network structural attributes [13, 18] as well as individual attributes [19, 20] in the adoption and dissemination of agricultural innovations, by looking at patterns in sets of practices rather than in one specific practice. Indeed, particularly in the case of CA principles, farmers view adoption as a series of diverse decisions, rather than a single decision [21]. Hence, Social Network Analysis (SNA) allows us to explore the characteristics of a network and understand emerging patterns within the innovation process. SNA is a powerful set of tools for representing and analyzing innovation dissemination and adoption [13, 22, 23] as well as interventions [24]. However, although many studies recognize the importance of analyzing attributes and learning relationships, most focus on one or the other, and ignore combined analyses that address both criteria jointly. In addition, studies that have used SNA in agriculture focus on the use of indicators such as density, intermediation, and centrality [14, 22, 25] and do not include individual attributes. An alternative way of analyzing both criteria is to use SNA techniques that include individual attributes such as age, education, and level of new practice adoption, as well as learning relationships.

Hence, there is need for more robust analysis when combining both structural and individual attributes to study the innovation process, in order to better understand how farmers share and adopt the CA principles and other complementary practices through their learning relationships [21, 26]. In agriculture, most social network analysis is concerned with the "one-mode" case—i.e., the analysis of relationship ties among peers. However, sometimes researchers study relations between two types of actors, where ties exist only between nodes belonging to different events— in this case, which farmers are practicing one or the other innovation. These kinds of data are referred to as "two-mode" networks, also known as affiliation networks or bipartite networks [27]. Thus, this paper aims to fill this gap by exploring the utility of performing SNA of affiliation networks, and then transforming the two-mode network into a one-mode network using the projection method, considering individual attributes as well as network indicators [28–30]. Then, using the two-mode network approach, we show how patterns emerge in the CA innovation adoption process by analyzing the learning relationships among smallholder maize farmers, while participating in the MasAgro[2] initiative in La Frailesca, a region in the state of Chiapas, Mexico.

---

[2]MasAgro, an initiative of Mexico's Ministry of Agriculture, Livestock, Rural Development, Fisheries and Food (SAGARPA) and the International Maize and Wheat Improvement Center (CIMMYT), brings together national and international organizations in partnership with innovative Mexican farmers to obtain higher and more stable crop yields (https://www.cimmyt.org/es/project-profile/masagro-productor/).

2 Methods

2.1 Data Collection and Management

Using a snowball sampling method, 222 maize farmers were surveyed to obtain information on relationships for learning CA practices in the La Frailesca region, Chiapas, Mexico. Farmers who participated in the Sustainable Modernization of Traditional Agriculture (MasAgro) initiative, implemented by CIMMYT during the 2013–2014 period, were selected. Data were collected using an egocentric approach to capture direct personal information from the stakeholders. Using a catalog of agricultural practices, the questionnaire included questions such as: Which practices you are performing? From whom did you learn to apply these practices? Other attributes were also recorded, such as age, education, municipality to which the farmers belong, area size, and yields. The present study assumes that if two farmers apply the same practice, this implies interaction. In order to analyze emerging patterns in the innovation process, a matrix was constructed based on the source of information. We also constructed three networks: (1) an affiliation network of actors and innovations; (2) a network of agents or farmers; and (3) a network of events or innovations. Finally, a vector based on an Innovation Adoption Index (*InAI*) was constructed in order to apply a regression model.

2.2 The Affiliation Network and Projection

The Affiliation Network

The affiliation network is based on a two-mode network [27], where the first mode represents the set of actors $N = \{n_1, n_2, \ldots n_g\}$ in this case the smallholder farmers, and the second mode represents the set of events $M = \{m_1, m_2, \ldots m_g\}$ or the agricultural innovations they practice. The affiliation network matrix is denoted by $A = \{a_{ik}\}$ and represents the affiliation of the actors with the events [28]. The element A takes the value 1, if actor i takes part in the event k, and 0 otherwise.

$$a_{ik} = \begin{cases} 1 & \text{if actor } i \text{ practice an innovation } k \\ 0 & \text{otherwise} \end{cases}$$

The sum of each row in the affiliation network matrix is the number of innovations an actor performs. The sum of each column represents the number of actors practicing a specific innovation. Actors and innovations can be also represented independently as a "one-mode" value network as follows.

Network of Actor Co-Memberships (Smallholder Farmers)

Using the one-mode network approach, the matrix denoted by X^N indicates the number of memberships shared by each pair of actors. There is a relationship between two actors if they share at least one event or agricultural practice. We then construct an adjacent matrix of this X^N network of size $n \times n$. An element of X^N is noted by x_{ik}^N with i, k $\in$ N. x_{ik}^N takes the value 1 if i and k have at least one event in common and if $i \neq k$, and takes the value 0 otherwise.

$$x_{ik}^N = \begin{cases} 1 & \text{if } i \text{ and } k \text{ share at least one agricultural practice and if } i \neq k \\ 0 & \text{otherwise} \end{cases}$$

We can then build the value matrix X_V. If, for example, two farmers share two different practices, then we can set the value of the coefficient $x_{ik} = 2$. The value matrix is

$$X_V^N = A \; x \; A^T$$

where A is the affiliation matrix and (A^T) the transposed matrix. The values on the diagonal of X_V^N represent the number of innovations each farmer applies.

Network of Events (Innovations)

We then use the same method to construct the innovations network, formed by a set of events or innovations M and edges L_M. Two innovations have an edge if at least one farmer applies the two innovations. The adjacency matrix of innovations X^M was constructed and noted as x_{jl}^M. A coefficient of this matrix is

$$x_{jl}^M = \begin{cases} 1 & \text{if } j \text{ and } l \text{ share at least one agricultural practice and if } j \neq l \\ 0 & \text{otherwise} \end{cases}$$

As with the matrix of farmers, we can build the valued matrix of innovations X_V^M. If two innovations have two farmers in common, then we can set the value of the coefficient $x_{jl}^M = 2$. This matrix is

$$X_V^M = A^T \; x \; A$$

with A the affiliation matrix and A^T its transposed. The values on the X_V^M diagonal represent the number of farmers per innovation.

2.3   Innovation Adoption Index

In order to measure the farmers' innovation levels, the *InAI* was constructed, which represents the proportion of innovations adopted by a particular farmer [31]:

$$InAI_i = \frac{\sum_{j=1}^{n} Innov_{i,j}}{n}$$

where $Innov_{i,j}$ = Adoption of the j innovation by farmer i. n = number of innovations performed by farmers in the study region.

2.4   Data Analysis

The data analysis was carried out in three stages: (1) actors and CA practices were analyzed to identify relationships and sources of information; (2) agricultural practices were analyzed to determine patterns of adoption; and (3) innovation adoption based on network and individual attributes were analyzed in order to explore the influence of network indicators and individual attributes in the innovation adoption process. First, the information source matrix was analyzed based on the affiliation network (Sect. 2.2). Second, centrality measures were calculated based on the three networks; then, using the network of innovations (network of events or innovations) and the Yule's Q coefficient [32] a heat map was created to identify emerging patterns in the innovation adoption. The *community detection based on edge betweenness, Newman-Girvan* [33] function, was also applied at this stage to find which practices are performed "together" by farmers. Finally, in order to assess the impact of the network's structural and individual attributes on the individual *InAI*, we carried out a Multiple Regression Quadratic Assignment Procedure (MR-QAP) test, which are permutation tests for multiple linear regression model coefficients, using data organized in square matrices of relatedness among n objects [34].

The analysis and visualization in stages (1) and (2) were carried out using the R software (Circlize package [35] and Igraph package [36]). Using Ucinet 6.627 [37] farmers' attributes (age, municipality, area, and education) were recorded and converted to one-mode networks, as were the regression calculations using the MR-QAP function.

3   Results and Discussion

3.1   Actors and CA Practices

Extension agents, relatives, and other farmers are the main sources of information and learning for smallholder farmers. It is also clear that different practices are learned from different actors (Fig. 1). This coincides with findings in other regions [38] in which the extension agents and fellow farmers were the most important sources of knowledge. The adoption of one or another practice is based on multiple

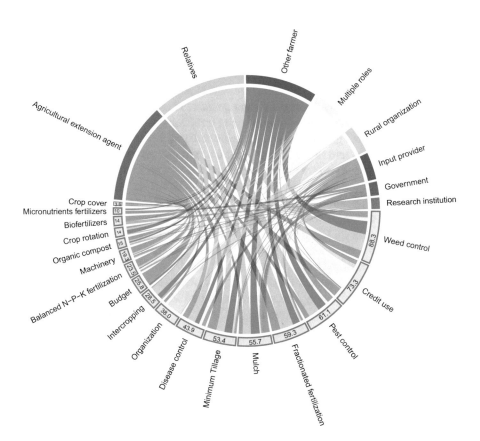

Fig. 1 Chord diagram of the one-mode CA practices network and information sources. *Note*: In the circle, the upper half (color) refers to different actors as information sources for learning, and the lower half (white) refers to CA and other complementary practices farmers are carrying out. Clockwise order (going from + to −) shows the importance of each component (sources and practices). Color lines indicate the "source" from which each "practice" is learned by smallholder farmers. The thickness of the line is weighted by the number times each farmer indicated different actors as the main source of learning. Numbers within the gray segments refer to the "network degree" as the number of farmers practicing that innovation in percentage terms

factors such as experience, the complexity of the practice or the infrastructure needed to apply new technologies. In this study, it is clear how smallholder farmers adopt the credit use from actors playing multiple roles—i.e., extension agents from public or research institutions and those working for private agriculture companies—because they are playing "brokerage" roles in the innovation process. In this sense, the links emerging from this kind of patterns are crucial for policy decision makers; for example, research institutions specially cannot always cover a large number of farmers in the innovation dissemination process and need other actors playing "brokerage" roles (see [39]) to spread the new technology or knowledge they are producing.

In the affiliation network, there were 222 farmers and 17 practices with a density of 0.358. The centrality results show that the most used practices in order of importance are those related to pest and disease management and nutrition; on the other hand, practices with low rates are those related to the use of specialized machinery, cover crops, micronutrients use, as well as, some of the most important practices in CA, such as crop rotation. The results indicate that, despite the value of CA practices being applied, there are some practices that have not been adopted by farmers, such as coverage with crops or green fertilizers. There are some contextual limitations to adopting CA practices, on the one hand, because of the difficulty of adapting the CA system to the context of small farmers; and on the other, because of the limited capacity to invest in aspects such as specialized machinery, which require a high initial investment [40]. In general, this pattern of relationships reflects that farmers apply practices that in the short term solve emerging problems like weed and pest control, which are crucial in maize production, and set aside those practices that represent CA principles, which in the medium and long term, would lead to higher and stable maize yields and increase productivity in long term. These results coincide with [41] about the limited adoption of CA practices in regions such as Central America and the Andean region, where socioeconomic constraints, access to suitable machinery, and uncoordinated efforts of stakeholders are factors that explain the low rates of adoption.

3.2 Pattern of Agricultural Practices

In the Yule's Q indicator, the index (-1 to $+1$ going from the red to a blue gradient) indicates a relationship between agricultural practices. A farmer who practices crop cover combines this with the use of specialized machinery (0.77), crop rotation (0.74) and micronutrient fertilizers (0.72); while farmers who access credit do not apply the specialized practices, like crop cover (-0.8) and crop rotation (0.05), required by conservation agriculture (Fig. 2a). The latter may possibly reflect that only capitalized farmers can implement the representative CA practices.

On the other hand, when the patterns of adoption of practices are analyzed under a "bundles" scheme rather than practices carried out at the individual level (Fig. 2b), three large groups are identified: the first (1) is related to the conventional pest and

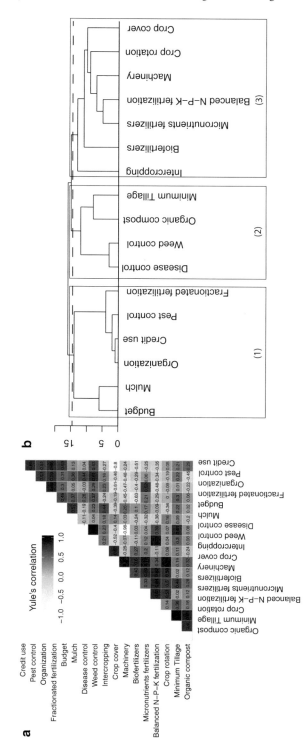

Fig. 2 (**a**) Correlation matrix heat map shows how each practice is adopted from each other, ranging from 1 (positive—blue) to −1 (negative—red), using Yule's Q correlation indicator. (**b**) A cluster analysis shows bundles of practices in the emerging adoption pattern. The height of the fusion on the vertical axis indicates the dissimilarity (using Euclidean distance) between two observations. The higher the distance is, the less similar the observations are

nutrition maize production practices (crop cover and pest control); the second group (2) includes practices related to minimum tillage, disease and weed control; and the third group (3) includes the most complex and costly practices that are carried out by a smaller group of farmers (19%) and promoted by the MasAgro initiative, such as crop rotation and the use of special machinery.

Adopting packages of agricultural practices or new technologies is a complex decision for farmers to make, because they involve several factors. On the one hand, factors related to the environmental conditions in which the production system developed, as well as the role of agricultural extension services, i.e., to follow up on and supervise the practices that are carried out. There are also some implicit individual attributes such as farmers' schooling, capitalization and access to specialized knowledge, particularly in the case of complex practices [21, 42]. Findings here agree with the study by [43] on the adoption of CA in Europe, where CA practices were adopted following a "step by step" strategy, where the largest farmers are the ones that adopt them more quickly, because they can absorb the risks involved in carrying out practices such as minimum tillage and mulching.

Nevertheless, the adoption of "bundles" of practices does not reflect a CA adoption pattern as it is conceived by extension services. Visualizing CA as a "package" may overestimate the decision-making process faced by the farmer, because it treats the adoption process as a simple decision, rather than as a series of interrelated decisions [21] in which synergies and complementarities arise. In Brazil, CA adoption was encouraged by the reduction in production costs and evident reduction in soil erosion [41]. However, particularly in southeastern Mexico and Central America, short-term priorities and land tenure restrictions are socioeconomic constraints to adopting CA, and sequential implementation is needed for smallholder farmers considering the management of local knowledge [44]. However, in other regions, findings suggest that it is not clear that CA is adopted through sequential implementation using the "ladder approach," and that possibly the "all in one" approach, going from small to larger scales, would probably be more feasible [10].

3.3 Innovation Adoption Based on Network and Individual Attributes

Table 1 shows the descriptive statistics of the main variables analyzed in the MR-QAP regression model that was constructed based on the network of events or innovations and the individual attributes network, modeling the effect of two sets of independent variables (network variables and individual attributes) and a dependent variable related to the *InAI*. Network variables such as *betweenness* capture the "power" of an actor based on his position in the network, that is, the higher the number of "powerful" actors in the network, the higher the number of shared innovations. The *innovation co-attendance* variable was constructed based on the

Table 1 Descriptive statistics and Pearson's correlation coefficients for network structural and individual attributes

Variables	Min	Max	Mean	Std dev	Pearson's correlation coefficients					
					1	2	3	4	5	6
Dependent variable										
1. lnAI	5.9	82.4	35.7	14.0						
Independent variables										
Network variables										
2. Betweenness	0	1	0.47	0.34	0.54***					
3. Innovation co-attendance	0	11	3.15	1.87	0.52***	0.40***				
Individual attributes										
4. Schooling	0	17	4.23	3.84	0.12**	0.02	0.12**			
5. Age	21	89	53.31	13.64	0.00	0.00	−0.07*	−0.05		
6. Same municipality	1	3	—	—	0.01***	−0.03**	0.01	0.02	−0.01	
7. Same area	0.5	34	3.75	3.92	0.01	0.00	0.01	0.02*	0.01	0.00

Note: Statistics for network variables calculated based on transformations of two-mode networks into one-mode affiliation network using UCINET 6.627 [37].
***$p < 0.001$, **$p < 0.01$, *$p < 0.05$

event or innovations matrix, and it assumes that if a farmer applies the same practice as another of his fellow farmers, they have a strong learning relationship. Therefore, they are likely to have the same innovation adoption pattern.

On the other hand, the individual attribute variables capture several of the farmers' characteristics, such as education, age, whether they belong to the same municipality, or whether they have a similar amount of land as the other farmers. This makes it possible to answer questions such as, for example, do farmers who live in the same municipality tend to use the same agricultural practices? This is crucial for agricultural extension services, whether public or private, because these types of patterns reveal the priorities or restrictions farmers face when adopting CA practices.

When analyzing the innovation adoption process, endogenous and exogenous effects that determine adoption can be identified [14]. The exogenous effects are related to specific characteristics, such as the education or age of the members of a network, that have a direct effect on the decision to adopt a practice or not. Endogenous effects are related to how a member of the network behaves at the individual level when taking a decision (for example, whether to adopt a new technology). Table 2 indicates that both network and individual attributes are moderately associated with the *InAI*. When analyzing the network indicators, the betweenness index in this case indicates the value of farmers who are carrying out multiple practices that others also perform [28]. This indicates that there are farmers who play the role of "brokers," are structurally well positioned within the network, and share the same practices as other farmers; this is reinforced when analyzing the same practice co-attendance indicator. The fact that a farmer applies the same practices as his peers implies that he has similar knowledge and experience about one or another practice. This explains two phenomena: first, farmers have been exposed to the same learning sources and similar experiences; and second, farmers who have similar knowledge reach similar processes of "understanding" as their peers. This would facilitate the flow of information and knowledge of the most

Table 2 MR-QAP regression model

Dependent variable: *InAI*	Standardized coefficient	*P*-value	Std err
Network attributes			
Betweenness	0.394	0.001	0.010
Same practice co-attendance	0.360	0.001	0.002
Individual attributes			
Academic level	0.076	0.001	0.000
Age	0.033	0.020	0.040
Municipality	−0.054	0.001	0.003
Surface	−0.013	0.001	0.002
Intercept	0.000	0.000	0.000
R-square (adjusted) (0.414)		0.000	

Note: MR-QAP linear regression method (2000 permutations.) using UCINET 6.627 for Windows [37]

specialized practices for research centers and/or agricultural extension services, and indicates the importance of promoting interactions between different actors, in order to promote the adoption of CA practices [45].

Finally, age and education were the most important individual attributes for reaching a higher innovation index. Education has been a variable that determines the adoption of new farming practices [19, 46, 47], given that farmers with a higher educational level are more likely to apply practices that require a higher specialization level; as for age, younger farmers always tend to assume higher risks [19, 48].

It is important to mention that these findings suggest association and not causality, and when interpreting the R-square, one should keep in mind that network variables may be overestimated due to the structure of the data [30]. However, this type of analysis allows studying emerging patterns in the structure of innovation adoption and highlights how analyzing the structure of relationships and individual attributes is important for exploring patterns of adoption behavior, especially for decision makers and agricultural extension agents.

Central America, which extends from southeastern Mexico to Panama and the Caribbean, is a region most vulnerable to disaster because of its location, its high climatic variability, as well as the extreme events that have occurred in recent years [49], and which continue to significantly affect the region's small farmers. Studies suggest that to improve food production, it would be much more efficient to implement agricultural practices such as CA because they would improve food security and mitigate climate change [50]. This could be achieved mainly by providing educational agricultural programs for farmers, so they could implement new practices and improve some of the ones they are already applying [44]. This case study shows that SNA is a tool that could contribute to a better understanding of the educational patterns and processes that emerge during interactions aimed at adopting farming practices that can increase the resilience of the region's agriculture to climate change.

4   Conclusions

Using SNA, this study examined the emerging patterns in the process of disseminating and adopting CA practices. We found that "bundles" of CA and other complementary practices emerge separately, as part of the interactions between the different stakeholders involved in the innovation process, but also as part of farmers' individual attributes and relationships. A key policy implication is that the government needs to strengthen participatory methods to promote sustainability in agriculture, rather than apply the usual hierarchical mechanisms in the innovation dissemination process. The main implication of our study for methodological issues is the use of SNA as a tool for analyzing innovation dissemination and adoption in sectors such as agriculture, and how SNA methods can reveal patterns in different analytical units/levels: the technology or practice, the individual, the external influences or the innovation system. A limitation of this kind of approach is that it privileges structural analysis over that the discursive, and some gaps could remain for a better understanding of CA adoption. This also indicates that future research needs to combine qualitative and quantitative methods when conducting SNA, in order to achieve a deeper understanding of discursive formations in the innovation dissemination and adoption process, mainly for the individual dimension.

References

1. SIAP: Anuario Estadístico de la Producción Agrícola. https://nube.siap.gob.mx/cierreagricola/ . Accessed 01 July 2017
2. Ranum, P., Peña-Rosas, J.P., Garcia-Casal, M.N.: Global maize production, utilization, and consumption. Ann. N. Y. Acad. Sci. **1312**, 105–112 (2014)
3. Shiferaw, B., Prasanna, B.M., Hellin, J., Bänziger, M.: Crops that feed the world 6. Past successes and future challenges to the role played by maize in global food security. Food Sec. **3**, 307–327 (2011)
4. Eakin, H., Perales, H., Appendini, K., Sweeney, S.: Selling maize in Mexico: the persistence of peasant farming in an era of global markets. Dev. Chang. **45**, 133–155 (2014)
5. Appendini, K.: Reconstructing the maize market in rural Mexico. J. Agrar. Chang. **14**, 1–25 (2014)
6. Pretty, J., Sutherland, W.J., Ashby, J., Auburn, J., Baulcombe, D., Bell, M., Bentley, J., Bickersteth, S., Brown, K., Burke, J., Campbell, H., Chen, K., Crowley, E., Crute, I., Dobbelaere, D., Edwards-Jones, G., Funes-Monzote, F., Godfray, H.C.J., Griffon, M., Gypmantisiri, P., Haddad, L., Halavatau, S., Herren, H., Holderness, M., Izac, A.-M., Jones, M., Koohafkan, P., Lal, R., Lang, T., McNeely, J., Mueller, A., Nisbett, N., Noble, A., Pingali, P., Pinto, Y., Rabbinge, R., Ravindranath, N.H., Rola, A., Roling, N., Sage, C., Settle, W., Sha, J.M., Shiming, L., Simons, T., Smith, P., Strzepeck, K., Swaine, H., Terry, E., Tomich, T.P., Toulmin, C., Trigo, E., Twomlow, S., Vis, J.K., Wilson, J., Pilgrim, S.: The top 100 questions of importance to the future of global agriculture. Int. J. Agric. Sustain. **8**, 219–236 (2010)
7. Pretty, J.: Agricultural sustainability: concepts, principles and evidence. Philos. Trans. R. Soc. Lond. Ser. B Biol. Sci. **363**, 447–465 (2008)
8. Power, A.G.: Ecosystem services and agriculture: tradeoffs and synergies. Philos. Trans. R. Soc. Lond. Ser. B Biol. Sci. **365**, 2959–2971 (2010)
9. Erenstein, O., Sayre, K., Wall, P., Hellin, J., Dixon, J.: Conservation agriculture in maize-and wheat-based systems in the (sub)tropics: lessons from adaptation initiatives in South Asia, Mexico, and Southern Africa. J. Sustain. Agric. **36**, 180–206 (2012)
10. Thierfelder, C., Baudron, F., Setimela, P., Nyagumbo, I., Mupangwa, W., Mhlanga, B., Lee, N., Gérard, B.: Complementary practices supporting conservation agriculture in southern Africa. A review. Agron. Sustain. Dev. **38**, 16 (2018)
11. Spielman, D.J., Ekboir, J., Davis, K.: The art and science of innovation systems inquiry: applications to Sub-Saharan African agriculture. Technol. Soc. **31**, 399–405 (2009)
12. Kilelu, C.W., Klerkx, L., Leeuwis, C.: Unravelling the role of innovation platforms in supporting co-evolution of innovation: contributions and tensions in a smallholder dairy development programme. Agric. Syst. **118**, 65–77 (2013)
13. Spielman, D.J., Davis, K., Negash, M., Ayele, G.: Rural innovation systems and networks: findings from a study of Ethiopian smallholders. Agric. Hum. Values. **28**, 195–212 (2011)
14. Matuschke, I., Qaim, M.: The impact of social networks on hybrid seed adoption in India. Agric. Econ. **40**, 493–505 (2009)

15. Basu, S., Leeuwis, C.: Understanding the rapid spread of System of Rice intensification (SRI) in Andhra Pradesh: exploring the building of support networks and media representation. Agric. Syst. **111**, 34–44 (2012)
16. Cadger, K., Quaicoo, A., Dawoe, E., Isaac, M.: Development interventions and agriculture adaptation: a social network analysis of farmer knowledge transfer in Ghana. Agriculture. **6**, 32 (2016)
17. Janker, J., Mann, S., Rist, S.: Social sustainability in agriculture – a system-based framework. J. Rural. Stud. **65**, 32 (2019)
18. Boahene, K., Snijders, T.A., Folmer, H.: An integrated socioeconomic analysis of innovation adoption. J. Policy Model. **21**, 167–184 (1999)
19. Feder, G., Umali, D.L.: The adoption of agricultural innovations. A review. Technol. Forecast. Soc. Chang. **43**, 215–239 (1993)
20. Mwaseba, D.L., Kaarhus, R., Johnsen, F.H., Mvena, Z.S.K., Mattee, A.Z.: Beyond adoption/rejection of agricultural innovations: empirical evidence from smallholder rice farmers in Tanzania. Outlook Agric. **35**, 263–272 (2006)
21. Ward, P.S., Bell, A.R., Droppelmann, K., Benton, T.G.: Early adoption of conservation agriculture practices: understanding partial compliance in programs with multiple adoption decisions. Land Use Policy. **70**, 27–37 (2018)
22. Díaz-José, J., Rendón-Medel, R., Govaerts, B., Aguilar-Ávila, J., Muñoz-Rodriguez, M.: Innovation diffusion in conservation Agriculture: a network approach. Eur. J. Dev. Res. **28**, 314–329 (2016)
23. Hall, A., Rasheed Sulaiman, V., Clark, N., Yoganand, B.: From measuring impact to learning institutional lessons: an innovation systems perspective on improving the management of international agricultural research. Agric. Syst. **78**, 213–241 (2003)
24. Valente, T.W.: Network interventions. Science (New York, N.Y.). **337**, 49–53 (2012)
25. Spielman, D.J., Davis, K., Negash, M., Gezahegn, A.: Rural innovation systems and networks: findings from a study of Ethiopian smallholders. Agric. Hum. Values. **28**, 195 (2011)
26. Khataza, R.R.B., Doole, G.J., Kragt, M.E., Hailu, A.: Information acquisition, learning and the adoption of conservation agriculture in Malawi: a discrete-time duration analysis. Technol. Forecast. Soc. Chang. **132**, 299–307 (2018)
27. Borgatti, S.P., Everett, M.G.: Network analysis of 2-mode data. Soc. Networks. **19**, 243–269 (1997)
28. Faust, K.: Centrality in affiliation networks. Soc. Networks. **19**, 157–191 (1997)
29. Everett, M.G.: Centrality and the dual-projection approach for two-mode social network data. Methodol. Innov. **9**, 205979911663066 (2016)
30. Baird, T.: Who speaks for the European border security industry? A network analysis. Eur. Secur. **26**, 37–58 (2017)
31. Muñoz-Rodríguez, M., Rendón-Medel, R., Aguilar-Ávila, J., García-Muñiz, J.G., Altamirano-Cárdenas, J.R.: Redes de Innovación: Un acercamiento a su identificación, análisis y gestión para el Desarrollo Rural. Fundación Produce de Michoacán/Universidad Autónoma Chapingo, Morelia (2004)
32. Lewis-Beck, M.S., Bryman, A., Futing Liao, T.: Yule's Q. http://methods.sagepub.com/reference/the-sage-encyclopedia-of-social-science-research-methods/n1097.xml (2004)
33. Girvan, M., Newman, M.E.J.: Community structure in social and biological networks. Proc. Natl. Acad. Sci. U. S. A. **99**, 7821–7826 (2002)
34. Borgatti, S.P., Everett, M.G., Johnson, J.C.: Analyzing social networks. SAGE, Los Angeles, CA (2013)
35. Gu, Z., Gu, L., Eils, R., Schlesner, M., Brors, B.: Circlize implements and enhances circular visualization in R. Bioinformatics. **30**, 2811–2812 (2014)
36. Csardi, G., Nepusz, T.: The igraph software package for complex network research. InterJournal Complex Syst. **1695**, 1 (2006)
37. Borgatti, S.P., Everett, M.G., Freeman, L.C.: Ucinet for Windows: Software for Social Network Analysis. Analytic Technologies, Harvard, MA (2002)

38. Opara, U.N.: Agricultural information sources used by farmers in Imo State, Nigeria. Inf. Dev. **24**, 289–295 (2008)
39. Klerkx, L., Aarts, N.: The interaction of multiple champions in orchestrating innovation networks: conflicts and complementarities. Technovation. **33**, 193–210 (2013)
40. Corbeels, M., de Graaff, J., Ndah, T.H., Penot, E., Baudron, F., Naudin, K., Andrieu, N., Chirat, G., Schuler, J., Nyagumbo, I., Rusinamhodzi, L., Traore, K., Mzoba, H.D., Adolwa, I.S.: Understanding the impact and adoption of conservation agriculture in Africa: a multi-scale analysis. Agric. Ecosyst. Environ. **187**, 155–170 (2014)
41. Speratti, A., Turmel, M.S., Calegari, A., Araujo-Junior, C.F., Violic, A., Wall, P., Govaerts, B.: Conservation agriculture in Latin America. In: Farooq, M., Siddique, K.H.M. (eds.) Conservation agriculture, pp. 391–415. Springer, Cham (2015)
42. Dhar, A.R., Islam, M.M., Jannat, A., Ahmed, J.U.: Adoption prospects and implication problems of practicing conservation agriculture in Bangladesh: a socioeconomic diagnosis. Soil Tillage Res. **176**, 77–84 (2018)
43. Lahmar, R.: Adoption of conservation agriculture in Europe. Lessons of the KASSA project. Land Use Policy. **27**, 4–10 (2010)
44. Cotler, H., Cuevas, M.L.: Adoption of soil conservation practices through knowledge governance: the Mexican experience. J. Soil Sci. Environ. Manag. **10**, 1–11 (2019)
45. Conley, T.G., Udry, C.R.: Learning about a new technology: pineapple in Ghana. Am. Econ. Rev. **100**, 35–69 (2010)
46. Abdulai, A., Huffman, W.E.: The diffusion of new agricultural technologies: the case of crossbred-cow technology in Tanzania. Am. J. Agric. Econ. **87**, 645–659 (2005)
47. Bordenave, J.D.: Communication of agricultural innovations in Latin America: the need for new models. Commun. Res. **3**, 135–154 (1976)
48. Warriner, G.K., Moul, T.M.: Kinship and personal communication network influences on the adoption of agriculture conservation technology. J. Rural. Stud. **8**, 279–291 (1992)
49. FAO: Disaster risk programme to strengthen resilience in the dry corridor in Central America. http://www.fao.org/emergencies/resources/documents/resources-detail/en/c/330164/ (2015)
50. Arcanjo, M.: Pioneering Agricultural Adaptation: Building Climate Resilience in Central America. Climate Institute, Washington (2018)

Mapping Informal Organization Through Urban Activism: The Case of Self-Organized Spaces in the City of Naples

Pasquale Napolitano, Pierluigi Vitale, and Rita Lisa Vella

Abstract The Naples associative dimension "from below" is considered to be a unicum at national and international level. This paper develops an exploratory research project looking at urban community hubs as social and political networks detectable by virtue of their organizational informality, exploiting the potential of data mining from social-media platforms in a user generated way. After a preliminary investigation to collect a list of active organizations in Naples, we have identified 32 organizations that have preserved some key elements over time: denomination, location, main actors. Then, we have verified and mapped the correspondence of their performed activities with the priority themes of the cultural production system defined by the European Urban Agenda. Therefore, we have built a dataset from Facebook events promoted by urban organizations and its participants; then, we have analyzed it through qualitative and quantitative data mining, and finally, we have visualized it and then proposed a cultural interpretation. The main expected results have been: firstly, to explore the proximity of the considered entities by comparison of the geographical and the social network space; secondly, to contribute to the scientific debate about the "relational ecosystem in the cities" both in theory and methods. The potential implication of this study could be considered in both fields of the interdisciplinarity of academic theories and methods, because of the proposal and use of both social sciences and statistical knowledge, and in the field of the policy strategy, because of the possibility of exploitation as a tool able to intercept the social capital of the territory and monitor the action of informal organizations operating in that territory.

P. Napolitano
Science and Technology for Information and Communication Society, Graphics, Vision, Multimedia IRISS/CNR, Naples, Italy

P. Vitale (✉)
Department of Political and Communication Sciences, University of Salerno, Fisciano (SA), Italy
e-mail: pvitale@unisa.it

R. L. Vella
Department of Business and Management, LUISS Guido Carli, Rome, Italy

© Springer Nature Switzerland AG 2020
G. Ragozini, M. P. Vitale (eds.), *Challenges in Social Network Research*,
Lecture Notes in Social Networks, https://doi.org/10.1007/978-3-030-31463-7_10

Keywords Urban networks · Social network analysis · Organizational informality · Community · Collective action

1 Introduction

It has been almost 50 years since the publication of Mitchell's *Social Networks in Urban Situations* [1] and the concept of "social network" is now acknowledged to be as important in the whole academic area of Social Studies. The field of Social Network Analysis is still developing in concepts and methods, above all with regard to the recent diffusion of social media and social networking websites [2]. Furthermore, cities and urban life are preferential framework to observe the contemporary social configurations for a better understanding of urban communities and of types, processes, and effects of social interaction [3]. In general, we would like to divide the discussion on urban networks into two main categories: the one of the World City Network paradigm referring to the inter-city relations and the overall connectivity among the cities as a consequence of globalization [4–6], and the one of the intra-city relations that looks at the communities and the networks inside the city [7–10] and at large the works of the sociologist Barry Wellman [11, 12] and Craven and Wellman [13]. The former includes the studies of city performance, above all in terms of economic and governance strategies for achieving efficiency goals; the latter is more focused on themes of participation, identity, sense of belonging, and creation of value. The same ongoing emphasis on the smart city question could be figured through these two big categories, depending on the focus on the network as infrastructure and technologies (with efficiency goals), or the focus on the network as social relations (with liveability and participation goals) [14–16].

Far from being so distinctly different, contemporary cities seem to be aimed at increasing new configurations of urban life where liveability and efficiency goals could be strongly connected and digital technologies and "bottom-up" initiatives will play a central role [17, 18]. Nowadays, within the city there are clusters of varied vocation, located in strong territorial proximity, a kind of community hub [19] based right on the informality of its activities, impacting the territories through forms of complementary action enabling the social capital of the territory. Following Latouche [20], the added value of this phenomenon is constituted precisely by its lack of organizational formalization and consequent adaptive dynamics of their capabilities. The contribution is to investigate a methodology of mapping the network of these inner-city organizations, by virtue of their organizational informality, exploiting the potential of data mining from social-media platforms in a user generated way. The basic assumption is that the city activists are not engaged in the creation of formal organization, hence the need for methods to capture the scattered social capital of every association by verifying their relations and their relational topic.

The research set is the City of Naples, in the South of Italy. One of the main reasons and interesting point of the city as research setting is the density and the complexity of Naples housing and meanings and, at the time, the full preservation of the city connotation over the years. This is also one of the main points of similarity with the neighborhoods of other big Mediterranean cities, despite the Neapolitan structural lack of services and infrastructure networks of European standard. In this sense, the Neapolitan neighborhood is the set of a continuous gamble staged by those who share the urban space, often not by choice, betting that the others' proximity should be positive for their own existence.

Therefore, we are talking about a part of the city where residents can (and should) act by themselves because of the atavistic absence of a well-framed administrative city project. Hence, by choosing Naples as case study, we could observe one of the particular cases in which the residents' engagement sometimes could be a much more effective solution from the point of view of the answer to the needs and the quality of life [21]. In other words, this urban malleability allows the neighborhood to be a place of "local economy," crafts, and both old and new knowledge, interpreting the need to give back to the city the role of a dialectical place of staging between individuals, crowds, and ethnic groups, leading to a possible coexistence despite differences.

In the first place, there are some preliminary clarifications about main concepts and main stages of the research strategy through a mixed methods research [22–25]: identification of urban organization; mapping (qualitative analysis); online activity exploration (qualitative analysis); Facebook data mining (scraping; quantitative analysis); interpretation (qualitative analysis).

The main expected results are: firstly, to explore the proximity of the considered entities by comparison of the geographical and the social network space; secondly, to contribute to the scientific debate about the "relational ecosystem in the cities" both in theory and methods. The potential implication of this study could be considered in both fields of the interdisciplinarity of academic theories and methods, because of the proposal and use of both social sciences and statistical knowledge, and in the field of the policy strategy, because of the possibility of exploitation as a tool able to intercept the social capital of the territory, to evaluate the action of informal organizations and their potential of self-organizational to meet the needs of one's own community by referring, at least from the point of view of some thematic areas, to cultural and creative supply chains.

2 Research Strategy and Method

For our purpose, we have followed the Johnson et al. [23] definition of mixed methods research: "Mixed methods research is the type of research in which a researcher or team of researchers combines elements of qualitative and quantitative research approaches (e. g. use of qualitative and quantitative viewpoints, data collection, analysis, inference techniques) for the broad purposes of breadth and depth of understanding and corroboration" (ibidem, p. 123). Mixed methods research (beyond this point, MMR) is common in different academic fields: Wisdom and

Creswell [24] and Borrego et al. [22] define and discuss the core characteristics of qualitative, quantitative and MMR; the former describe the expansion of MMR from the origin in social sciences to the health and medical sciences; the latter talk about the MMR in engineering education research. Nowadays, several references [17, 18, 26] agree about the importance to study the urban context, as one of the main parts of the general social context, with a combination of a deep qualitative understanding with valid and generalizable findings or relationships among variables. For this reason, we believe that this approach would be particularly suitable to studying the city combining qualitative understanding with offline and online data collection and analysis. Moreover, the article tries to analyze the urban network of the persons who attend the events promoted by some locally based organizations, by means of data-mining procedures performed through social media sites focusing on Facebook in order to answer to the lack of MMR on this platform and to the explicit recommendation [27] for future research in that sense.

The starting point of the project of the mixed methods research of this study is the observation[1] [28] of the popular neighborhoods of Naples, a big city in Mediterranean Europe, and in particular the phenomenon of the adaptive re-use of large abandoned buildings. The aim of this first stage is to identify active urban organizations that have preserved some key elements over time (denomination, location, main actors).

To have a list of organizations, we refer to:

- Social Centers (SC) features, already recognized as best practices in the activity of urban decision-making (see [26]);
- exploratory analysis already carried out on the territory [29].[2]

Following Vittoria and Napolitano [29] we look at the Neapolitan Social Centers as informal communities, "creative places" and drivers of urban productivity. Here, we are referring to "informal organizations" as those organizations with free participation of actors who are able to develop enduring trustful relationships that may be governed by entrenched moral economies organized on particular sets of values. In fact, the configurations assumed by the Social Centers in the different cities, their success in terms of *imitation* [30], and the considerable diffusion of the initiatives in the wake of collective and informal decision-making contribute to open the debate on the interpretation of their economic and social effectiveness and on the possible fostering of this particular type of *urban development laboratory* [31].

About the informal communities operating in the area of the Municipality of Naples, there are some key goals for an analytical modality:

---

[1] Following Cardano [28], observation is "the methodological choice of qualitative research to meet a general need of the whole field of social research: the need to cope with the complexity of the investigated phenomena" (ibidem, p. 7, original source in Italian).

[2] Vittoria and Napolitano [29] focus on the contribution of the Neapolitan social centers to the processes of urban governance, in order to propose a synoptic reading of the phenomenon starting from the various cultural and creative sectors in which the activities of each social center are located.

- to understand how the network of citizen activists is distributed;
- to intercept the mobility of activists between the different organizations;
- to understand the criterion and purposes of mobility.

Our analysis focuses on all the SCs of the observation area connected to process of greater awareness in terms of civicness and collective action for the city [32, 33].[3] On the other hand, the overall vision of SC citizens is still lacking. In order to provide a contribution in this regard, the present study uses digital community detection tools to produce a social listening action on the examined social entities.

Until now, this was above all an issue of sociological interest, in order to consider firstly the coexistence of such networked forms of organization with markets and hierarchies throughout the history of modern economic institutions, without automatically prompting for positing the existence of an alternative mode of production [30]. Secondly, the sociological interest was related to the field of multiple motivations for work, emphasized by the classic Hawthorne studies and the connected tradition of Human Relations management (cf. [36]). In detail, the literature on knowledge work and knowledge management has been particularly adamant in suggesting that monetary motivations are insufficient for understanding or motivating contemporary knowledge work where "creativity" has become a core feature. Similarly, non-monetary motivations have been a precondition [37] and a significant driver of economic action before and during the history of modern capitalism [38].

Later, the advent of the digital age opened up the public realm, the commons, the sharing of goods and services, and infrastructure within urban areas. In doing so, this blurred the lines where the formal economy ends and the informal economy begins. Avenues of public information stimulated improved governance in addressing specific issues within the cities, enhancing citizen engagement in decision-making and greater governmental accountability [39]. Moreover, the UrBes [40] benchmarks study (2015) reveals the rising importance of the non-profit sector in the social fabric, confirmed by the positive evolution of the most of the indicators on institutions, volunteering, social cooperatives, and the paid employment in the social sector. During 2015 in Naples, the growth of non-profit sector—as volunteering, solidarity aims, disadvantaged people involvement—is directly related to the fall of average citizen's income.

Recently, unlike the general volunteering context, a process designed to include informal organizations activities in the citizen governance project has been triggered. The road is that of a "shared administration," understood as co-participation in the management of public space, beyond the traditional set of fixed roles: the one who designs the space (institutional people) and the one who uses it (common

---

[3]Empirical studies on different urban context highlighted the importance of these social formations for anthropological and sociological analysis: from the studies on the SCs role as bearers of new codes and symbols in language [34] to the new practices of digital communication introduced by the strategies of collective action, recently observed in the case of Macao in Milan [35].

citizens).[4] The process runs, thanks to two driving forces: on the one hand, the national and international debate [21, 26, 35, 41, 42]; on the other hand, the process at several levels of new practices of urban development and social innovation, launched by groups that seem to be the Neapolitan form of *informal and creative communities* [43, 44], outposts of new sociality and spreadable planning that test new practices through collaborative systems.[5] Consequently, the specific context requires to identify tools for monitoring organizations in terms of communication and framing among activists.

Some economists hold that institutions are as important to development as economic geography and trade [45]. Most of the research on institutions and economic development compare institutions of different countries, contrasting constitutions and legal frameworks, the structure and performance of governments, and the politics of decision-making and interest-groups [46]. These forces shape labor markets, entrepreneurship, and the growth of firms [47]; on these bases, recent studies are investigating regional economic and civic networks; Feldman and Zoller [48] quantify interregional differences in brokers and deal makers, such as venture capitalists, as a force in shaping the geography of knowledge-intensive industries. According to additional research on the connections between business networks and civic networks (i.e., NGOs, community organizations, churches, etc.), the strength of these connections shapes the direction and the strength of regional adaptation to external shocks [49, 50]. Further studies argue that regional adaptation is a function of both the strength of major groups and the quality of the informal connections or formal coalitions, where both the values of coalescence and competition are important to regional dynamism [51]. In addition, by studying the processes of economic specialization of particular places, other scientists have found that successful adjustment—or resilience—depends on the local ability to capture activities that can become the basis of new regional economic specializations [52].

Although there are various ways that this can occur from region to region, it is now well known "why" some places adjust better than others. In regard to these problems, some researches adopt methodological choices that reverse the common horizontal perspective in defining the models of development of an area. The awareness of the deep differences in acting collectively, indeed, explains the need to conduct in-depth investigations and selective collection of best practices.

---

[4]In 2011, first time in Italy, the legal category of "Common Good" was included in the "purposes and fundamental values" of the Municipality's Statute creating a specific administrative department: Resolution 446/2016, art. 3 establishes "the Municipality, in order to protect future generations too, guarantee full recognition of common goods as operational for the exercise of fundamental human rights in the ecological framework." http://www.comune.napoli.it/flex/cm/pages/ServeBLOB.php/L/IT/IDPagina/16783.

[5]In 2016, ECF was engaged in a mapping exercise of best practices that culminated in the publication of a Magazine compiling 26 civil-public partnerships based on principles of the commons.

In order to face the recent EU Urban Agenda priority themes,[6] most of the above-mentioned studies merge on the analysis of the formation processes of the informal social networks and on the analysis of the subsequent collective actions.

The final list includes 32 organizations.[7] They are all open collectives, strongly characterized by the urban political and social context and to some extent linked to the life of the neighborhood in which they are located.

Then, we follow the stages of our MMR in an ongoing combination of qualitative and quantitative procedures with data collection and analysis in order to:

- map the geographical distribution and purposes of organizations (Fig. 1),
- map the cluster of the activities carried out by the Social Centers on the grounds of the correspondence of their performed activities with the priority themes of the cultural production system defined by the European Urban Agenda [26] (Figs. 2 and 3),
- build a dataset pertaining Facebook events promoted by selected urban organizations and its participants; analyzing it to obtain descriptive statistics about the number of participants, distribution of networks and clusters, mobility of activists, etc.; visualizing it for a better understanding of the criterion and purposes of the mobility (Fig. 5).

3   Data Mining and Visualization

Digital Methods (DM) consist in a set "techniques for the study of societal change and cultural condition with online data" [53]. The overarching methodological scope of DM is to "follow the medium," that is, to conceive the Internet as an environment where "native" methods of research are built into online devices, such as Google Web Search and Facebook's Graph Search. Therefore, DM urge researchers to take advantage of the "natural logic the Internet applies to itself in gathering, ordering and analyzing data" [54]—as with tags, links, hashtags, or retweets. DM evaluates the nature and affordances of the digital environments by studying these structure communications and interactions among social actors as well as how these can be used as methodological strategies and techniques by social researchers.

The science of social network analysis has co-evolved with the development of online environments and computer-mediated communication. Unique and precise data available from computer and information systems have allowed network scientists to explore novel social phenomena and to develop new methods. Additionally, advances in the structural analysis and visualization of computer-mediated social networks have informed developers and shaped the design of social media tools [55].

---

[6]Specifically, we are talking about the 12 priorities established as the central themes of the EU's smart, sustainable, and inclusive growth strategies for 2020. (EU Ministers responsible for Urban Matters, 2016) cf.: http://urbanagendaforthe.eu/pactofamsterdam/.

[7]Twenty-five engaged in activities of cultural production and 7 with housing purposes.

1. L'Asilo	13. Lido Pola	25. Terzopiano Autogestito
2. Damm	14. Zero81	26. Ex Asl Abitata
3. Ex Opg	15. Aula FLEX Orientale	27. Ex Scuola Schipa
4. Officina 99	16. Aula Autogestita R5	28. Annona
5. Scugnizzo Liberato	17. Laboratorio S.K.A.	29. Villa De Luca
6. Santa Fede Liberata	18. Mensa Occupata	30. CROSS
7. Giardino Liberato Materdei	19. Spazio di Massa	31. T.N.T.
8. Laboratorio Insurgencia	20. Biblioteca Gramsci-Dax	32. Zia Ada
9. Selva della Lacandona	21. Aula C4	
10. Villa Medusa	22. DADA	
11. Il Gridas	23. Mezzocannone Occupato	
12. Nuova Casa del Popolo Ponticelli	24. AULA LP Lettere Precarie	

Fig. 1 Organizations geo-distribution and purposes in the City of Naples

Exhibition, Meeting, Entertainment, Fairs	Handcraft	Design	Film, Video, Radio, Tv	Music	Books and Newspaper	Museums, Libraries, Archives, Management of hystorical sites	Videogames Softwares, Hacking
DAMM	DAMM	Ex Opg	DAMM	Ex Opg	DAMM	DAMM	Terzopiano Autogestito
Ex Opg	Ex Opg	Mensa Occupata	Il Gridas	L'Asilo	Ex Opg	DAX	
Giardino Liberato	Giardino Liberato	S.K.A.	L'asilo	Officina 99	L'Asilo	Giardino Liberato	
L'Asilo	Il Gridas	Scugnizzo Liberato	Nuova Casa del Popolo	S.K.A.	Mezzocannone Occupato	L'Asilo	
Lido Pola	L'Asilo	Selva della Lacandona	Santa Fede Liberata	Santa Fede Liberata	Villa Medusa	Lido Pola	
Officina 99	S.K.A.	Terzopiano Autogestito	Terzopiano Autogestito	Scugnizzo Liberato		Mezzocannone Occupato	
S.K.A.	Selva della Lacandona		Villa Medusa Occupata	Terzopiano Autogestito		Nuova Casa del Popolo	
Santa Fede Liberata			Zero81	Zero81		Santa Fede Liberata	
Scugnizzo Liberato						Villa Medusa	
Zero81							

Fig. 2 The activities of the U/S classification (the organizations classified according to the U/S classification)

The first step, before the social media data collection, was a preliminary investigation to get a list of active organizations in the reference area (Fig. 1).

The samples in Fig. 1 display the activities thickening in the historic city center (20 out of 32 cases). The phenomenon could be linked to the availability of

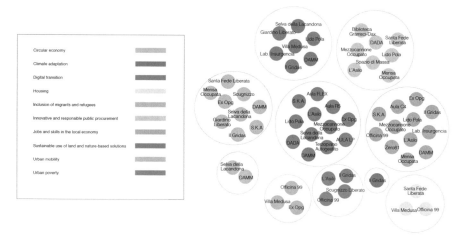

Fig. 3 Organizations grouped by development categories. Critical tasks—Source: our elaboration from different sources

the so-called *sleeping giants* [56], vacant buildings with a high symbolic value.[8] Evidence shows interesting results in locations that collect clusters of different activities: informal organizations proliferate where the formal dimension—like cultural associations—already exists along with activities of private stakeholders in cultural and creative sectors, from the perspective of subsidiarity between the different actors involved in the various sectors of community welfare.

Finally, in the cases examined, professional uncoded practices are developed in synergy with forms of social demand, focusing foremost on an operational dimension of social acting, instead of an analytic one.

The choice to observe the nature of the activities carried out by the SCs derives from the emphasis that in some models of urban development analysis has been attributed to informal communities in improving urban productivity. By virtue of this analysis and of the considerable difficulties in terms of empirical observation, where in many cases it is difficult to recognize a structured production process, the research oriented the following stages of observation towards something materially offered by these actors. In the specific case of the SCs, it could be useful to identify the first capacity in the nature of political action, which often translates itself into services offered to citizenship. Looking directly at the offer of cultural products/services, for example, it is possible to rediscover initiatives (events, debates, shows, exhibitions) that would not have been produced, if not

---

[8]Ten out of 20 cases are linked to the occupation of academic spaces. The occupations date back to the early stage of SCs activity (until 1995). Then, some students involved in that experiences preserved their political and social engagement out of the academia context.

within the SCs reality. Thus, we develop a first visual cluster of the activities carried out by the Social Centers (Fig. 2). Therefore, we will try to grasp in the alignment between the actions–situations of the centers with the priorities established in the supranational context.

The figure shows in columns, entitled to the individual U/S sub-sectors, the names of the SC which correspond to them, by the nature of the activities carried out. Overall, the Neapolitan SCs tend to perform activities that correspond to 8 of the 10 categories of the cultural production system [26].[9] Looking at the broader ability to establish relations with local institutions, SCs action, as forms of cooperation *from the bottom*, goes in the direction of *civil-public partnerships,* as defined by the European Urban Agenda through 12 priority themes of development, based on participatory principles able to provide out-of-box solutions and to promote the strengthening of democratic legitimacy [26]. Within this development system, cultural production is a fundamental resource that emerges as a driving force of success in urban strategies, and, first and foremost, as a key element of urban attractiveness, and also as a catalyst for differentiation, revitalization, and change. Furthermore, in order to support a balanced economy and a social development, active citizen involvement is essential to allow greater social inclusion and a close association between culture and civil society. It is precisely on the basis of the role that the cultural supply chains play within the European Urban Agenda, that it seemed to us effective to use the 12 guidelines for urban development categorization in order to have a picture of which lines are activated in the city area by the considered informal organizations (Fig. 3).

What emerges is an extremely lively and varied picture, in which 10 of the 12 lines of development examined were put in place by at least one of the analyzed organizations.[10] Instead, coming to the central part of the analysis, the graph

---

[9]There is a greater inclination towards activities classified as: "artistic representations, entertainment, conferences." Equally significant is the remarkable representation of sectors such as publishing, theater, music, and crafts, sectors that are historically and largely present within the city's productive fabric and that in these spaces are looking for new production methods (theater, publishing) or the recovery and innovation of practices that, both for the absence of services provided by institutional sources and for a gentrification phenomenon, are moving away from the historical center and the popular neighborhoods (crafts, music). However, despite this growing awareness, urban heritage is still threatened. Although most historic neighborhoods still have a central function, uncontrolled urban development jeopardizes their preservation. Some neighborhoods show clear signs of decay, while others are commodified for tourist activities. Cultural tourism has sometimes resulted in further deterioration or in the exclusion of local populations and in a loss of authenticity due to over-branding. In many cities, gentrification processes in central areas resulted in the eviction of low-income populations into informal settlements [21].

[10]Specifically, for the Neapolitan territory there are two particular lines of development in which the organizations are more committed: the first one is "Jobs and skills into local economy," which refers to the activities of local economies through learning and production paths; the second one is "Digital Transition," which includes all the activities of facilitating access to digital media. There are also many examples of the search for alternative ways of governance of the public heritage (Innovative and responsible public procurement) and examples of the search of Nature Based

(Fig. 5) is carried out on a dataset pertaining to Facebook events promoted by urban organizations and its participants.

Starting by the previous mapping [29], we have selected the informal community engaged in the field of cultural production field and for each one of these organizations, we have collected the official Facebook page ID, a unique number that identifies only a single page on Facebook, used subsequently as key of research in the collection phase. With this type of process, we have avoided any case of homonymy or pages managed unofficially with the same or similar name.

With a custom python script, we have made a query to the Public Facebook API, collecting all the events made by each organization in a dataset. Then, in order to collect just the (technically declaring) participants at the events of each organization, we decided to make our queries just to the "event attending" endpoint, excluding from the data all the people that answered "no" or "don't know." In this way, since from the beginning we have focused our data just on the people active in these communities.

The dataset contains 11,712 rows, starting from the first day 01/01/2017 until the final day of data collection, 01/03/2017. The first 136 rows are the list of the events in the period in exam. The remaining large part is a kind of edge list, in which there are the users and the events they have indicated to take part on Facebook. For each one of these events, as attribute, we have taken all the reference of organizations that organized them.

Starting by this tabular dataset, we have built an *adjacency matrix* for a *two mode network* [57], in which we decided to observe the relationship between participants and organizations as nodes, with the first as a source and the latter as a target (Fig. 4).

For the visualization of this directed bipartite graph, we used the software Gephi (gephi.org) in which we have chosen to assign two different colors, light blue and blue, respectively, to participants and organizations. Nodes match if a person has been at events promoted by the organization during the 3 months we have examined.

Sizes are based on the *in-degree* [58, 59], so the bigger nodes are the organizations with more participation at the events. As layout for our visualization we have adopted the Fruchterman Reingold layout [60], because it seems to be the best choice to highlight the state of public and communities.

In this network configuration the position of nodes participants and nodes organization is based on the density of the relationship.

It is possible to see some nodes uniformly positioned at the margins of the graph, the ones only related with one organization-node (as a loyal community); group of nodes equally connected with two or more nodes; many nodes positioned in the

---

Solutions, which is easy to explain with the need to rationalize and regenerate a very high density urban space, a space totally devoid of public green, and at the same time, the need to reconstruct ecological chains interrupted for decades by urban and rural fabric. To a lesser extent, but with significant experience, there are actions related to housing, inclusion and assistance to migrants, and also the fight against poverty and alternative forms of mobility are represented.

Data Collection
3 levels of data collection

	Facebook API	
Level 1	Level 2	Level 3
Previous Mapping	**Events list**	**"Event Attending"**
Organizations (and its Facebook official page)	Date-Time Event ID Title N° of participants	Name Unique User ID
↓		↓
Nodes		**Nodes**

Fig. 4 Data collection phases from the Facebook API

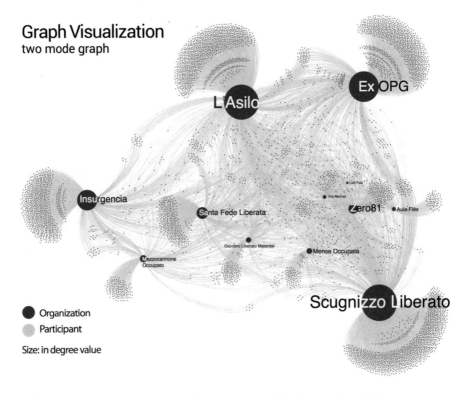

Graph Visualization
two mode graph

● Organization
● Participant

Size: in degree value

Fig. 5 Graph visualization: network of organizations and attendees to the activities

mid of the graph, that represent the unpolarized public, with a participation divided between many events promoted by the different organizations in the network.

4 Evidence

Nowadays, the shape of the cities is not just structured around the role of classical civic organizations, but citizens could lead the process of creation and sharing new forms of communication, aggregation, and participation in urban life. Indeed, the gradual hybridization of digital-physical practices in the cities divides itself into apocalyptic and integrated people interested in proving or rejecting the wins or the losses of this process.[11] This contribute is focused on active urban networks not only because of the proposed interpretation of the new role of citizen participation due to the digital-physical urban practices, but also because of the account of the dynamism[12] of the urban life. In fact, the visualization in the previous paragraph is not interpreted as a static classification of the city life but as a specific picture of a moment that will change over time.

Firstly, in the urban dynamics every citizen is a *prosumer* in so far as he contributes in "market and design information vital for the production process [...] share data, information, and knowledge" [63]. In the case of the present research, we are considering two kinds of data:

- Facebook Events created by activists and shared in order to call for physical participation.
- The Answers to this call, by clicking on "will participate" or "maybe participate,"[13] that is a public promotion of intent with or without the next stage of physical involvement.

According to that, "prosuming" is primarily due to the associations networks (or their activists), who "design and share information, data and knowledge vital for the production process" that will carry on and take place elsewhere. In that stage, the attendees are just responding to their call in structured and expected forms: by clicking a Facebook button and sharing. Here, we found one of the main shortcomings of a social media approach to the city networks. But, since we are

---

[11]Groups of academics are engaged in debunking myths and extreme generalization using both quantitative and qualitative methods, for example, the Senseable City Lab of MIT is proposing a human-centric approach to shift from the research on urban network to the research on active urban network [18].

[12]Dynamism of urban life could refer to frequent change of residence [61]; "dynamic areas, i.e., coherent geographic areas shaped through social activities" [62]. Here, dynamism is referred to the mobility of citizens in participating to the activities of different social centers and to the consideration that the network analyzed is not a static representation of the citizens' networks due to the speed and frequency of online-participation changes.

[13]In order to restrict bias, we have only considered the "will participate" option.

considering self-organized spaces in the city, citizens are far from being just users of a top-down artifact. In fact, there are not any completely isolated nodes: in some way, everyone is a member of the urban community with different interests degree.

Understanding participation is one of the main goals of media studies [64]. The reference is not so unusual even if we are observing a city and not a brand or a media company or a cultural product in the sense of media studies. The original proposal of this work is that today, more than ever, the hybridization of top-down and bottom-up dynamics suggests placing the city on a scale from "highly branded" to "highly participated" [3]. In the former, there is a majority of top-down processes lead by public administration with economic giants like IBM [65] as new players, in the latter, there is a majority of bottom-up processes where citizens are primarily responsible for the definition of information flows and decision-making processes. As a matter of fact, looking at the graph (Fig. 5), it is possible to classify participation dynamics in networks of "core-attendees," those ones with nodes exclusively linked to one of the main associations, and networks of "cross-attendees," with nodes linked to two or more associations. It is not coincidence that the main networks of "core-attendees," Scugnizzo Liberato, L'asilo, Ex OPG, Insurgencia, are located at the extremes of the represented network and, such as compass point, these nodes are useful to navigate it (Fig. 6). Furthermore, it is not coincidence that these associations are the ones born in districts with tradition of social and political activism, which they have been able to collect and reshape or remix with the new purposes and values of their specific identity. They rely on a pre-existing stronger collective identity.

At first glance, this would appear as a representation of communities and publics [66], where the former are the networks of "core-attendees," with main associations anchored in local interactions and social cohesion, and the latter are the network of "cross-attendees" (Fig. 7), linked by unstructured and transitory social bonds with no intention of participating in the process of creation and negotiation of one collective identity or a sense of belonging. But, since the reason for the publics is more visibility than identity, this interpretation could be a response to the possibility of overrated participation due to the habit of public promoting the attendance (by clicking a button) without a consequent physical attendance at the event. However, this classification would be useful from the SCs point of view, in a "branded" sense of the informal organization to investigate the loyalty of affiliated ones, but it would underestimate the value of participation in the city life.

On the contrary, sporadic participation to the activities of one or more associations could reflect a weak affiliation to the association community, but a strong involvement into the urban one. In that sense, the networks of "cross-attendees" are particularly interesting to our aims and to future research:

- they are not isolated nodes or an inactive public, but nodes who participate in the initiatives of more than one association and, in this way, they are the only connection between them;

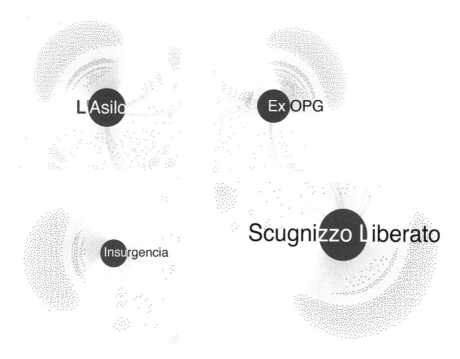

Fig. 6 Zoom on networks of core-attendees participating in the activities of one specific organization

– they are "highly participated," with the possibility to intercept bottom-up process where citizens are primarily responsible for the definition of information flows and decision-making processes.

Thus, the nature of community is changing into a complex, stratified networks system inclusive of different sub-networks, local and global bounds, personal and social purposes. Therefore, it is possible the conceptualization of community as a social network with local (as in traditional neighborhoods) or global bounds, as in internet-based communities: "rather than relating to one group, people live and work in multiple sets of overlapped relationships, cycling among different networks" [11].[14]

Hence, here we face one of the main problems about participation, meaning the trend towards static interpretation, where attending is something out from the creation process, and activists and attendees are stuck in their roles forever. Actually, one of the main points about detecting and interpreting urban networks

---

[14]About networks and communities: Bessant [7], Alvarez et al. [8], Giuffre [9], Rainie and Wellman [10], and at large the works of the sociologist Wellman [12] and Craven and Wellman [13].

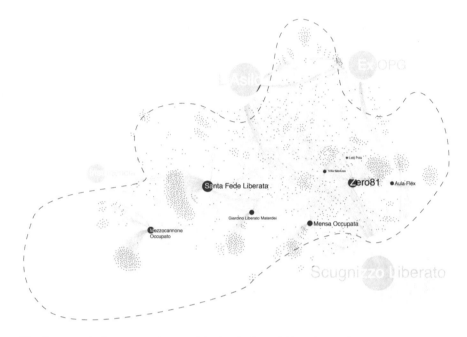

Fig. 7 Network of cross-attendees participating in the activities of more than one organization

structure is that mapping the interactions is not like building some kind of pyramidal structure. Understanding the process complexity means not only to give up on the understanding of crowd-based values, tiny and sporadic values by definition, but also on driving force of the adaptive dynamics of new economic and social forms. Moreover, recalling the premises about urban network studies, for those interested in intercepting and supporting the "smartness" of cities [14–16], this study supports the interpretation of the network as social relationships, where liveability and participation goals are already drivers of efficiency. So, the smartness seems even more connected to the lack of organization formalization and to the atypical logics beyond the management of policy powers, by activating the social capital across powerful human networks.

The organizational structure emerging by the networks is based on the hybrid links between associations and citizens that enable considerations about the difference between collective and connective action. The impact of digital media is stronger on the second one dynamics, thanks to logics of personalization for self-expression or self-validation. But, the important thing is the possibility of co-occurrence of the two logics in what Bennett and Segeberg [67] call DNA—*digital networked action*—where the lack of strict standards for collective identity frames is what provides the possibility of scaling up quickly, tracking moving political targets and bringing up different issues. Finally, the specific status of Naples form of living reveals itself as a unique thing in the middle of that scale that goes from HB (highly branded) to HP (highly participated),[15] where every day there are processes of hybridization between top-down and bottom-up practices and every actor (or node) is able to capitalize the widespread socio-cultural value and launch a collective action.

In conclusion, intercepting and enhancing the whole structure of urban participation would allow the policymaker strategies to benefit of the multi-layered social capital in different levels of effort and to develop a multi-narrative that transcends the strictly economic logic in order to intercept social, cultural, and identity elements.

5 Conclusions

Summarizing what was observed in the previous sections, it is possible to highlight some conclusions:

- All these initiatives are influencing the renewal of our cities from bottom-up. These regeneration policies have directly contributed to sustainable urban development agendas. By reactivating existing structures and developing new urban services, the adaptive re-use of the existing buildings has helped to contain urban sprawl and foster density. It has also contributed to the maintenance of local businesses and infrastructure in inner city areas.
- The ability to unlock the rich cultural heritage value of these areas might be a decisive factor for forging meaningful urban transformations. A more concerted engagement with the multiple dimensions of urban informality, and the ways they intersect with notions of memory, identity, and the "right to the city," is therefore a pivotal thing [39]. Cultural systems embedded in informal settlements should be recognized, analyzed, and reinvested in order to stimulate community-based development and deepen our understanding of the ways in which these multicultural spaces have engendered resource-sharing [66], starting right from human resources, represented by the changing magma of activists.
- These social media visualizations provide an insight into that collective imagination that continuously informs and is reproduced in the communications that communities engage in. In communities that have a physical co-presence, this work of imagining community [70] is also undertaken through the organization of events and occasions that enable the network to come together and experience itself as a community, and by the day to day effective "labor" of the processing community. In this sense, these "communities" are "recursive communities" [71]: communities that have, as an important purpose, the ongoing reproduction of

---

[15]This concept is deeply linked to that of the *performativity of language*: the approach to performative language, introduced by Austin [68] with the definition of performative utterance, proposes to no longer look at language as a descriptive tool, but to understand it properly as an act. Ritual phrases from credible people in appropriate contexts do not describe action but do it [69].

themselves as communities. Rather, the sense of solidarity and community is anchored in the practices. It is by doing things together that participants share something and belong to something more meaningful, it is the common practice that connects a multiplicity of individual projects and outlooks into a common movement. People frequent these spaces (either online or offline) to engage in a particular practice and it is by engaging in such practice that they connect together and form social bonds.

– In general, we conclude that the investigated informal communities enter into the "dead" urban system to vitalize it. Evidence shows that the local social centers' actions are moving to face the main EU Urban Agenda priority themes. Thus, the emerging social centers' critical ability is in building a bridge between their own (political) motivations/action strategies and the wider EU policy framework.

From the data in our possession, the question remains open whether this participation translates into an impact on the territory, and if the activity of the most—even more numerous—self-identified communities translates into a greater social impact than more open communities, or if this greater cohesion is limited to an act of belonging.

The originality of this paper lies in the lack of prior similar research. At the same time, there is a limit due to the impossibility to replicate the research because of the new privacy policy of the social media Facebook. However, the methodological approach could be replicated in different contexts, wherever possible.

References

1. Mitchell, J.C. (ed.): Social Networks in Urban Situations. Analyses of Personal Relationships in Central African Towns. Manchester University Press, Manchester (1969)
2. Scott, J.: Social Network Analysis. SAGE, London (2017)
3. Vella: La città, il digitale, l'everyday life: un approccio transdisciplinare. I casi Napoli e Barcellona. Unpublished doctoral dissertation. Department of Communication and Political Science, University of Salerno (2018)
4. Khanna, P.: Connectography: Mapping the Future of Global Civilization. Random House, New York (2016)
5. Taylor, P.J., Derudder, B.: World City Network. A Global Urban Analysis. Routledge, London (2004)
6. Beaverstock, J.V., et al.: Attending to the world: competition, cooperation and connectivity in the World City network. Global Netw. 2(2), 111–132 (2002)
7. Bessant, K.C.: The Relational Fabric of Community. Palgrave Macmillan, New York (2018)
8. Alvarez, L., Borsi, K., Rodrigues, L.: The role of social network analysis on participation and placemaking. Sustain. Cities Soc. 28, 118–126 (2017)
9. Giuffre, K.: Communities and Networks. Using Social Network Analysis to Rethink Urban and Community Studies. Polity Press, Cambridge (2013)
10. Rainie, L., Wellman, B.: Networked: The New Social Operating System. MIT Press, Cambridge (2012)
11. Wellman, B., Boase, J., Chen, W.: The networked nature of community: online and offline. IT Soc. 1(1), 151–165 (2002)

12. Wellman, B. (ed.): Networks in the Global Village. Life in Contemporary Communities. Routledge, New York (1998)
13. Craven, P., Wellman, B.: The network city. Sociol. Inq. **43**(3–4), 57–88 (1973)
14. Caird, S.P., Hallett, S.H.: Towards evaluation design for smart city development. J. Urban Des. **24**, 188 (2019)
15. Ratti, C.: Smart City, Smart Citizens. Egea, Milan (2014)
16. Hollands, R.G.: Will the real smart city please stand up? Intelligent, progressive or entrepreneurial? City. **12**(3), 303–320 (2008)
17. Iaconesi, S., Persico, O.: Digital Urban Acupuncture. Human Ecosystems and the Life of Cities in the Age of Communication, Information and Knowledge. Springer, Cham (2017)
18. Ratti, C., Claudel, M.: The City of Tomorrow: Sensors, Networks, Hackers, and the Future of Urban Life. Yale University Press, New Haven (2016). Resolution n. 446/2016
19. Lancichinetti, A., et al.: Characterizing the community structure of complex networks. PLoS One. **5**(8), e11976 (2010)
20. Latouche, S.: Il pianeta dei naufraghi. Bollati Boringhieri, Turin (1993)
21. Culture Action Europe: Polis and the People: Looking into the Future of Urban Cultural Policies. Culture Action Europe, Brussels (2017)
22. Borrego, M., Douglas, E.P., Amelink, C.T.: Quantitative, qualitative and mixed research methods in engineering education. J. Eng. Educ. **98**(1), 53–66 (2009)
23. Johnson, R.B., Onwuegbuzie, A.J., Turner, L.A.: Toward a definition of mixed methods research. J. Mixed Methods Res. **1**(2), 112–133 (2007)
24. Wisdom, J., Creswell, J.W.: Mixed Methods: Integrating Quantitative and Qualitative Data Collection and Analysis While Studying Patient-Centered Medical Home Models. Agency for Healthcare Research and Quality, Rockville (2013). February 2013, AHRQ Publication No.13-0028-EF
25. Domínguez, S., Hollstein, B. (eds.): Mixed Methods Social Networks Research: Design and Applications, vol. 36. Cambridge University Press, Cambridge (2014)
26. European Cultural Foundation: Build the City. How People are Changing Their Cities. European Cultural Foundation, Amsterdam (2016)
27. Highfield, T., Leaver, T.: A methodology for mapping Instagram hashtags. First Monday. **20**, 1–5 (2015). https://firstmonday.org/article/view/5563/4195
28. Cardano, M.: La ricerca qualitativa. Il Mulino, Bologna (2011)
29. Vittoria M. P., Napolitano P. (2017), Informal community as "creative places" and urban productivity drivers. The case of the Social Centers in Naples, in: Rivista Economica del Mezzogiorno, Bologna, Il Mulino, 1.
30. Granovetter, M.: Economic action and social structure: the problem of embeddedness. Am. J. Sociol. **91**(3), 481–510 (1985)
31. Persico, P.: Identità e Sviluppo. Laveglia Editore, Salerno (1997)
32. Rossi, U.: Lo Spazio Conteso. Il Centro Storico di Napoli tra Coalizioni e Conflitti. Alfredo Guida Editore, Napoli (2009)
33. Dines, N.: Tuff City. Berghahn Books, New York (2012)
34. Melucci, A.: An end to social movements? Introductory paper to the sessions on "new movements and change in organizational forms". Soc. Sci. Inf. **23**(4–5), 819–835 (1984)
35. Murru, M.F., Cossu, A.: Macao prima e oltre i social media. La creazione dell'inatteso come logica di mobilitazione. Stud. Cult. **12**(3), 353–371 (2015)
36. Wren, D.: The History of Management Thought. Wiley, London (2005)
37. Parsons, T.: The professions and social structure. Soc. Forces. **17**(4), 457–467 (1939)
38. Polanyi, K.: The Great Transformation: (The Political and Economic Origin of Our Time). Beacon Press, Boston (1957)
39. UNESCO: Culture: Urban Future; Global Report on Culture for Sustainable Urban Development. UNESCO, Paris (2016)
40. ISTAT/Comune di Napoli: Rapporto UrBes 2015. Il Benessere Equo e Sostenibile nelle Città (2015)

41. D'Ovidio, M., Cossu, A.: Culture is reclaiming the creative city: the case of Macao in Milan, Italy. City Cult. Soc. **8**, 7 (2016). https://doi.org/10.1016/j.ccs.2016.04.001
42. Micciarelli, G.: Le teorie dei Beni Comuni al banco di prova del diritto. La soglia di un nuovo immaginario istituzionale. Poli. Soc. **1**, 123–142 (2014)
43. Manzini, E.: Resilient systems and cosmopolitan localism—the emerging scenario of the small, local, open and connected space. Economy of Sufficiency, p. 70 (2013)
44. Manzini, E.: Design, When Everybody Designs. MIT University Press, Cambridge (2015)
45. Rodrik, D., Subramanian, A., Trebbi, F.: Institutions rule: the primacy of institutions over geography and integration in economic development. J. Econ. Growth. **9**(2), 131–165 (2004)
46. North, D.C.: Institutions, Institutional Change, and Economic Performance. Cambridge University Press, Cambridge (1990)
47. Acemoglu, D., Johnson, S., Robinson, J.: Institutions as the fundamental cause of economic growth. NBER n. 1048 (2004)
48. Feldman, M., Zoller, T.D.: Dealmakers in place: social capital connections in entrepreneurial economies. Reg. Stud. **46**(1), 23–37 (2012)
49. Benner, C., Pastor, M.: Whither resilient regions? Equity, growth and community. J. Urban Aff. **38**, 5 (2014)
50. Safford, S.: Why the Gardern Club Couldn't Save the Youngstown: The Transformation of the Rustbelt. Harvard University Press, Cambridge (2009)
51. Rodriquez-Pose, A., Storper, M.: Better rules or stronger communities? On the social foundations of economic change and its economic effects. Economic Geography. **82**, 1: 1–1:25 (2005)
52. Storper, M., et al.: The Rise and Fall of Urban Economies. Lessons from San Francisco and Los Angeles. Stanford University Press, Stanford (2015)
53. Rogers, R.: Digital methods for web research. In: Scott, R.A., Kosslyn, S. (eds.) Emerging Trends in the Behavioral and Social Sciences, pp. 1–22. Wiley, Hoboken (2015)
54. Caliandro, A.: Ethnography in digital spaces: ethnography of virtual worlds, netnography, and digital ethnography. In: Denny, R., Sunderland, P. (eds.) Handbook of Anthropology in Business, pp. 658–680. Left Coast Press, Walnut Creek (2014)
55. Rosen, D., Barnett, G.A., Kim, J.H.: Social networks and online environments: when science and practice co-evolve. Soc. Netw. Anal. Min. **1**(1), 27–42 (2011)
56. URBACT III: 2nd Chance—waking up the sleeping giants (2016). http://urbact.eu/sites/default/files/2nd_chance_state_of_the_art.pdf
57. Zaveršnik M., Batagelj, V., Mrvar, A.: Analysis and visualization of 2-mode networks. In: Meeting of Young Statisticians (2001)
58. Wasserman, S., Faust, K.: Social Network Analysis: Methods and Applications, vol. 8. Cambridge University Press, New York (1994)
59. Tsai, W.: Knowledge transfer in intraorganizational networks: effects of network position and absorptive capacity on business unit innovation and performance. Academy of Management Journal. **44**(5), 996 (2001)
60. Battista, G.D., Eades, P., Tamassia, R., Tollis, I.G.: Graph Drawing: Algorithms for the Visualization of Graphs. Prentice Hall PTR, Upper Saddle River (1998)
61. Jayapalan, N.: Urban Sociology. Atlantic Publishers, New Delhi (2002)
62. Gkatziaki, V., et al.: DynamiCITY: Revealing city dynamics from citizens social media broadcasts. Inf. Syst. **71**, 90–102 (2017)
63. Toffler, A.: Powershift: Knowledge, Wealth, and Violence at the Edge of 21st Century. Bantam, New York (1990)
64. Jenkins, H., Ford, S., Greene, J.E.: Spreadable Media: Creating Value and Meaning in a Networked Culture. New York University Press, New York (2013)
65. IBM: Smarter Cities Series: A Foundation for Understanding IBM Smarter Cities. https://www.redbooks.ibm.com/Redbooks.nsf/7abb875e4606df5785257a390065e0b7/24e5a12b8c58f8f98525783700706323 (2011)
66. Arvidsson, A., Caliandro, A.: Brand public. J. Consum. Res. **42**, 727 (2016)

67. Bennett, W.L., Segerberg, A.: The logic of connective action. Inf. Commun. Soc. **15**(5), 739–768 (2012)
68. Austin, J.L.: How to Do Things with Words. Oxford University Press, London (1962)
69. Austin, J.L.: Other minds. Proc. Aristot. Soc. **20**, 44 (1946)
70. Anderson, B.: Imagined Communities: Reflections on the Origin and Spread of Nationalism. Verso, London (1988)
71. Kelty, C.: Two Bits: The Cultural Significance of Free Software. Duke University Press, Durham (2008)

The Paths of the Italian Young Poets.
A Social Network Analysis
of the Contemporary Poetic Field

Sabrina Pedrini and Cristiano Felaco

Abstract This paper explores the social dynamics and the interactions between young poets and the Italian poetic network. More specifically, 70 poets were questioned about their participation in poetic activities and their cultural path. Using Social Network Analysis approach, we obtained preliminary information about the roles and structure of the Italian poetic network. Moreover, the explorative multidimensional analysis of a two-mode network was conducted using Multiple Correspondence Analysis in order to focus on the nature of relational data and the presence of asymmetry between actors and events in the network. The main results show low density network with many isolated experiences, where traditional publishers are still determinant in the building of personal career path and in the distribution network's support. Moreover, in a market that tends to "squeeze out," platforms and online journals are strongly emerging as places of co-production and co-distribution of the young poetry in Italy.

Keywords Italian poetry · Cultural network · Cultural market · Social Network Analysis · Multiple Correspondence Analysis

1 Introduction

Cultural production takes place in highly competitive sectors, but as Bourdieu [1] pointed out, some of these are narrow, niche, and with a low productivity level, pure market such as the one represented by poetic production. The careers of young artists, so uncertain, develop among logics linked to the ability to produce a culturally relevant product and to meet the market [2].

S. Pedrini (✉)
University of Bologna Alma Mater Studiorum, Bologna, Italy
e-mail: sabrina.pedrini@unibo.it

C. Felaco
University of Naples Federico II, Naples, Italy
e-mail: cristiano.felaco@unina.it

© Springer Nature Switzerland AG 2020
G. Ragozini, M. P. Vitale (eds.), *Challenges in Social Network Research*,
Lecture Notes in Social Networks, https://doi.org/10.1007/978-3-030-31463-7_11

In Italy, even if poetry is the source of all literary forms [3], poets have lost their role and their social prestige. This is a mass schooling's effect and disintermediation's phenomenon result due to which artists have ceased to be intellectual reference figures. The removal from these contents also derives from the fact that cultural institutions have ceased to be interested in youth productions. Moreover, the characteristics of the literary market and, within, the poetry's niche make it a pure market [4]: an area in which producers are also consumers and where the figure of a poet pure, damned and lonely does not exist anymore. Circularity is an organizational characteristic of high symbolic content's markets [5], where conventions rules define the process's stages, from artistic creation and distribution to consumption [6], which contents should be proposed, how these should be represented, appreciated, evaluated and what rules should regulate the relationship between artist and audiences [7]. Elitist and not popular [2, 8], this field has limited access and disclosure.

The not historicized poetic production remains "at the margins of the margins" and it almost never reaches the general public as it is in a society not prepared for poetic language, neither by the school, nor by the cultural system, nor by the information and communication's means. Therefore, those who practice this form of art rarely fall within a circuit or an institutionalized system but organize themselves through forms of transversal support: the young poets, in fact, have shown the need to "create a network," thanks to the fact that they perform side jobs in the publishing sector: all those who have this access are able to log the network and social networks allow to benefit to cultural sources, even for a simple *serendipity* process [9]. This sector, more than others, lives, thanks to informal networks' creation, which represents the problems related to the creation and maintenance of marketing and distribution activities solution. But something is changing: without going through traditional channels, the use of social networks is expanding the number of producers and with it the number of readers, making shorter the path towards institutionalization and fostering quicker access to those who, in the market, can favor the transformation from symbolic to economic capital.

In general, sociological and economic literature gave little space to this niche. This research aims to grasp the relational behavior between young Italian poets and, specifically, the relational processes that are established between peers and participation in different types of events (exhibitions, publications, etc.). To do this, 70 authors were selected: young, with an already started career, established [10], and their professional relationships network was built, thanks to Social Network Analysis (SNA) tools [11–13].

Most cultural production fields are studied through the approach of SNA [14–19]. In our research, the network structure of poetical field is formalized as a two-mode network in which poets are involved in different events such as festivals, lives, reviews, and so on. It permits to operationalize field positions through the network positions of poets and their affiliation in co-participations at the same events. Along with this, a Multiple Correspondence Analysis (MCA) was carried out to better outline similar patterns of participation of the poets interviewed [20]. The main results show a common situation to the rest of Europe [10, 16]: the atomization, the emergence of small cohesive subgroups that perform activities mainly online, in contexts where fixed costs can be drastically reduced; the lack of trust in their work and the importance that this can have in a context of transmission and construction of cultural capital; little interest and closure towards the dominant thought and consequent lack of attention to the received reviews [10].

This paper is structured as follows: Sect. 2 describes the context of poetry in contemporary Italy and circumscribes the sector in literature environment, Sect. 3 frames the research design that illustrates the methods and procedures adopted; Sect. 4 presents the results of the analysis concerning both to the whole network and to the participations to the single events; it concludes showing the results of the relationships between poets and events by means of MCA. The last section concludes with some remarks about the undertaken work.

2 Italian Poetic Context

Contemporary poetry is considered a minor literary genre: economic marginality[1] has been widely studied and it is notoriously marked by strongly elitist cultural markets [2, 23, 24]. After the late 1960s the figure of the poet has lost his social role along with his status. The reasons for this loss and isolation are many and profound: it is likely that the crisis of the literary genres, which came indeed in the 1960s, had largely damaged the popularity and visibility of poetry itself that, in the traditional ranking, was undoubtedly more appreciated.

As Mazzoni states [25], poetry is the most self-centered literary form to express private and singular experiences, not comparable to others. Because of this, it is also the most fragile form among the "weak arts" which are, in other words, all the literary arts fallen into crisis during the 1970s because of mass education and mass freedom of speech. The definition of poetry as a "weak art" underlines its lack of affinity with communication. However, being much related to spoken words, poetry is indeed similar to a non-art form [26]. Between the 1960s and the 1980s new media like television, for instance, have gained a leading role encouraging the growth of different art forms in which the dimension of communication in terms of knowledge [27] is more dominant [28].

Ultimately, starting from the late 1980s the institutional contest consistently weakened, in other words, universities, publishers, and newspapers lost interest

---

[1]The estimation of Italian sales of books in 2014 was 20% less compared to 2009. Poetry's sales are 0.6% of the whole literary industry. On the one hand, a history editor as Crocetti states that his sale is significantly decreased, while on the other hand Einaudi states more positive data. Inside the broader literary market, the micro-market of poetry is organized as an oligopoly where important editors (15% of the sector with over 50 publications per year) show how investing in poetry is profitable in the long run, thanks to the republishing of old authors [21]. These big authors are surrounded by small and even smaller authors (less than 10 publications per year) which are, however, very strongly specialized [22].

in those still not established contemporary poets [10]. The educational approach to poetry, especially in Italy, has remained in the hands of those who are highly educated and already master this knowledge operating in the literary industry, such as academicals, professors and teachers, literary critics and editors (as confirmed by the present study).

The field of literature nowadays is constantly changing. Poetry is changing, opening to mass-culture and mid-culture [29].[2] Internet along with new editorial spaces such as blogs, web magazines, and social networks is offering a whole new level for publishing and broadcasting allowing a different relationship with the readers to come through, more direct and without barriers. Barriers standing at the gates of the traditional publishing industry have fallen apart, leaving space for new leading characters to be born that would not have reached otherwise the chance to be published by important editorial brands [30]. Today, web and social networking are the most important instruments to cultural sources' access also through simple process of *serendipity* [9]: these allow to keep a strong motivation for those who carry out a job in culture, despite financial insecurity and risks connected to work in a context where the failure rate is high. To reduce the contests' risks, some authors practice the so-called side-line jobs where they fill in different positions in the editorial industry increasing gift economy [31], in which generosity and willingness are a great motivation for the artistic exchange, fundamental characteristic of all the members of this field. Subsequently they can also become "gatekeepers," involved in redacting, organizing events, publishing, reviewing, teaching, and mentoring. Translating these activities into the Bourdieu's forms of capital [30], not only they increase the artists' reputation helping them to develop their level of reputational capital, but also give them more visibility and recognition in the industry improving indeed their symbolic capital as a source of income for the economic one. The safeguarding of these activities even though difficulties characterizes them as internal goods [32].

Consequently, the community represents a rewarding and comforting structure where there is no institution capable of recognizing the work. In the Italian context we can talk more about small *meaning* communities rather than real currents of thought: the new relationships, which cannot be considered true friendships, are instead linked to a broader concept of *sociability* [33]. This strong and cohesive structure is typical in small markets with low profits made in the long run: the network must be extremely efficient and harmonious, embedded [34, 35], and loyal, where it is possible to develop a sense of community that puts the value of poetry at its core, guided by a faith along with continuous participation to the field, the one that Bourdieu calls *illusio* [36], which again improves the dynamics of prestige [4].

Poetry production, individual or collective, develops from poetic worlds [7], "autarkical world," "little world," "ghetto" [37], an organized, coherent, partly institutionalized space, a democratic society [38] that Anheier, Gerhards, and

---

[2]Forms of mass-culture that allow the public to have easy access and use along with a mid-culture represented by new entertaining and modern products.

Romo [39] call "archipelagos" inside of which dynamics of cooperation and group activities grow in an open but independent exchange: a space of relations [7, 40] and a "pure" market [40]. Poetry production can be also developed inside the cultural production field (former's subfield) through multiple subjects which have different positions and dynamics in the field, defined as that social space where relationships happen between classes of agents who fill different positions, a dynamic space where roles are not fixed and relations are *eo ipso,* hierarchically structured [30] not reducible to the individual agents' sum linked by simple relations or interactions, or, more precisely, cooperation [4, 41]. But today the separation between author "for the page" that follows esthetics and symbolic principle and author "for the stage" is not so clear anymore: the SNA allows us to see that the opening of a new market space (online journal) expanding the producers' audience will lead to the consumers arena increase and therefore to the sector's survival.

3 Methods and Procedures

In a first phase of the research,[3] to draw up a list of poets, in the absence of an exhaustive and official list of Italian poets, it has been decided to invite a group of experts linked to the Pordenone Legge event, one of the most important collective events in Italy. The sampling strategy was carried out accordingly to an *event-based approach*, i.e., through the selection of the actors on the basis of common characteristics, such as participation in the same event or belonging to the same group [42]. All the poets have been contacted through Facebook and asked to give their availability to be tested. This availability shows how informal contact and the use of social networks facilitate the initial relationship between subjects who do not know each other although have "mutual friends" whose value is recognized through a single contact, transmitting the sense of reciprocal trust [43]. The subjects were interviewed through a questionnaire concerning their curricula: publishers with whom they worked along with online journals, participation at festivals or readings, higher profile professional writers with whom they are in contact, educational qualification, and prevalent work activity. Data are entered in an incidence matrix in which the poets are placed in rows ($n = 70$) and events in columns ($n = 296$) and each link denotes the participation of that actor in that particular event: the relationship between poets and their co-participations can be represented by a two-mode (or affiliation) network [13]. As we have said, there is not an exhaustive list of poets to work on and most importantly it is difficult to identify the "professional" poet as it is a fluid figure which escapes categorization and whose quality transcends the concept of "Professionalism" as traditionally intended. Given the elusive nature of the phenomenon, it was preferred to conduct an exploratory study to investigate the contexts in which poets participate (festivals, collaborations

---

[3]The survey was carried out in August 2012.

with publishing houses, etc.) and what kind of relationship they maintain among themselves.[4] Attention was focused on the core-periphery structure with the intent to detect the network connectivity. We adopted the discrete approach of [45] that is based on the comparison of the adjacency matrix of the network. This method is computed by using a numerical method that finds a partition of nodes which is as close as possible to an idealized matrix; moreover, it allows to identify which actors and events in the core and which belong in the periphery and which events belong in the core and which events belong in the periphery. In the case of two-mode network, the core is a given partition of poets connected to many events belonging to a given partition of events and, at the same time, this partition of events is composed of events which are connected to many poets who belong to the partition of poets taken into account. Therefore, the set of poets in the core is defined by the set of events in the core and conversely [46]. In addition, degree centrality and eigenvector centrality for two-mode networks were used. These indices are usually used to study affiliation networks [47]: the degree centrality of a poet indicates the number of events with which it is affiliated, and the degree centrality of an event is the number of poets affiliated with it [48, 49]. Therefore, this measure reveals the level of activity or the number of contacts of the poets, as well as it traces the most important and widespread events in the sector (according to the size of their memberships) [50, 51]. Eigenvector centrality shows the duality of actor and event centralities [47] according to which actors' centralities are proportional to the centralities of the events to which they belong, and events' centralities are proportional to the centralities of their members [48].

In order to simultaneously observe actors and the events in the network, we used an alternative approach to analyze affiliation data without converting to co-affiliations which directly consider two-mode network. It performs a joint analysis of the two modes, thanks to the non-dyadic characteristic of two-mode networks [20]. Moreover, it contributes to an increasing understanding than the sole analysis of one-mode data, since much information in the original bipartite structure is lost because of the projection process [52, 53]. One way to jointly analyze actors and events is to apply factorial methods [13, 14, 54]. In our research, we adopted a Correspondence Analysis (CA) [55–57] for two main reasons: it is best suited to both the qualitative data used and to the explorative purposes of our research because it is not constrained by strict model assumptions or distributional requirements [53].[5] This method allows to treat the affiliation matrix as a two-way contingency table; more specifically, we use Multiple Correspondence Analysis (MCA), a variant of CA, that treats the affiliation matrix as a case-variable-variable matrix. MCA represents the relative frequencies within a contingency table as

---

[4]The analyses of core-periphery and centralities were performed through the software Ucinet 6 [44].

[5]CA of the affiliation network incorporates a similar relationship between actor and event as expressed in eigenvector centrality. In fact, the use of eigenvector centrality is coherent with that of factorial methods for the analysis of this kind of networks [13, 47].

the distance between individual row and column profiles and the distance to the average row and column profile. This option represents an effective approach to analyze and graphically represent the relationship between actors and events in a low-dimensional space [20, 58]. Moreover, MCA lets to understand the nature of relational data and the presence of asymmetry between poets and events in the network [13, 19, 20, 59, 60]; afterwards, it also permits to identify patterns of participation in the poetry field.[6]

4   Results and Discussion

The database includes 50 men and 20 women living in Italy, born between 1968 and 1988, with a bachelor's or master's degree. All of them have written in online journals and have at least one career collection: in particular, 81.4% of them have published more than two books, 22.9% have more than five publications in online journals, and 77% did not keep the received reviews. The interviewed poets often use social media, in particular, Facebook, for the dissemination of information about the field of poetry (82%), about one in four has created a literary blog on which it proposes its own compositions. It is interesting to underline that no interviewees declare that they earn their living from their artistic work; there is no place-based community that supports the artists in Italy, in fact, all carry out another activity for their livelihood. Few declare that they remember all the events they took part in, and they give little importance to them because they recognize they receive little attention and are aware of their work's low impact, even if they feel the need and continue to produce it. For the same reason, disillusionment, only a few keep a detailed press of reviews received [61]. However, interviewees common element is the need to have good relationships in order to publish their artistic works: more than half of them, to the question "Who is the most important person for you, professionally?" quoted the family, confirming that cultural activity in Italy is sustained by families not only in terms of trust and moral support, but above all, in terms of economic investment. Collaborative circuits and friendships are, indeed, crucial for sector and careers' development [62], and the literature sector economy has shown how increasing peer interactions creates innovation and creative growth [63].

In the next sections, we show the characteristics of affiliation network, more specifically the relationship patterns and co-participations of poets through MCA.

---

[6]The analysis was performed through the software Spad 5.5.

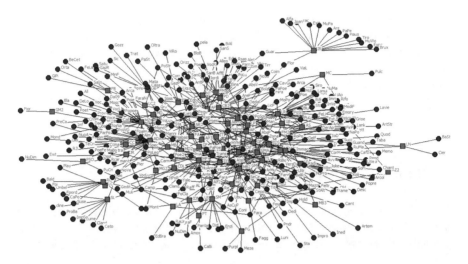

Fig. 1 Poets and events' affiliation network (squares represent poets and circles represent events)

4.1 The Poets' Affiliation Network

The graph that illustrates the two-mode affiliation matrix does not present isolates for which each actor is connected with an event just once (Fig. 1). The graph reveals the bridging role of some poets and events. Even if involved in many activities, some poets are marginal compared to the structure of the network[7]: IB, MC, LN, MZ2, MB3, FG, PT, LA, SL, LC, RC, PSO, GM2. Their structure is similar: all these authors are connected to events in which they are the only ones to take part. In fact, there are 210 out of 296 events that appear only once in the network. Therefore, events can be local or carried out abroad; publishers can be very small because in this sector many works are self-produced; as far as the online work often occurs on personal pages or blogs. A result so wide shows that many poets follow the "art for art's" rule, not pursuing a path towards institutionalization or notoriety.

The analysis of the centrality allows us to detect those overlapping memberships that foster the links' understanding between actors and events.[8]

As Table 1 shows, the authors who reach the highest positions in D are also among the first in the E: they are poets "for the paper" that prefer the writing activity to performances (only two public events appear in these lists). Every one of them is involved in the most important and influential events. That clearly

---

[7]The graphs linking poets/live events, poets/review received, poets/on line journals, poet/traditional publisher will be showed below and will be described in detail the individual positions identified.

[8]Thanks to the calculation of Degree Centrality (D). For the most influential we have used the Eigenvector Degree (E). For space requirements only the 10 poets with higher D and E are shown. The same for events.

Table 1 Poets and events with higher degree centrality and eigenvector centrality

Poets	Value	Events	Value
Degree centrality			
MZ—Matteo Zattoni	0.111	NaInd—Nazione Indiana	0.333
LR—Lidia Riviello	0.105	Lieto—Lietocolle	0.242
ADA—Azzurra D'agostino	0.101	Atel—Atelier	0.242
AR—Alessandro Raveggi	0.084	Dif—D'if	0.212
PD—Patrizia Dughero	0.068	Croc—Crocetti Perrone Ed	0.197
AP2—Alberto Pellegatta	0.064	AbsP—Absolute Poetry	0.182
FM2—Francesca Matteoni	0.061	Mond—Mondadori	0.182
MF—Matteo Fantuzzi	0.054	MYM—Marcos Y Marcos	0.167
CB2—Cristina Babino	0.054	NuArg—Nuovi Argomenti	0.136
GP—Gilda Policastro	0.054	Poes—Poesia	0.121
Eigenvector centrality			
AR—Alessandro Raveggi	0.336	NaInd—Nazione Indiana	0.457
LR—Lidia Riviello	0.319	AbsP—Absolute poetry	0.268
FM2—Francesca Matteoni	0.258	Croc—Crocetti Perrone Ed	0.266
ADA—Azzurra D'agostino	0.249	Dif—d'if	0.241
MS—Marco Simonelli	0.239	Lieto—Lietocolle	0.230
GP—Gilda Policastro	0.234	Atel—Atelier	0.198
CB2—Cristina Babino	0.219	MYM—Marcos y Marcos	0.178
MF—Matteo Fantuzzi	0.211	Trans—Transeuropa	0.163
MZ—Matteo Zattoni	0.199	Zona	0.160
AP2—Alberto Pellegatta	0.194	PaPoe—Parco poesia (Riccione)	0.149

emerges from high values of centrality measures: on the other hand, highlight the greater importance, within the network, of the online magazines that are today the main sources from which publishers draw information about newcomers or young authors. The results in this case show a strong coherence: Nazione Indiana[9] and Absolute Poetry[10] are considered the places to find potential editorial successes also in the narrative field.

The role of the festivals/events is interesting as, even if they are important in the interviews carried out, they do not turn out as central: Parco Poesia (12th)

---

[9]Nazione Indiana is a cultural project founded in March 2003 by an Italian group of writers, critics, and artists, with the purpose of literary promotion, which takes the form of an online magazine. The site's name was suggested by the poet Antonio Moresco to recall the particular sense of community among the native people of North America. With the aim of giving voice to texts and ideas that do not find their space in commercial publishing and information press, Nazione Indiana has published literary and critical texts, political and cultural initiatives, scientific divulgation and art exhibitions. It is considered today, in the context of narrative production too, the place on which editors of the most important publishing houses scout.

[10]Collective structure similar to Nazione Indiana, headed by different subjects, always belonging (in a lateral way) to the world of publishing, they carry out a series of projects including a festival in Monfalcone.

and Romapoesia Festival (13th) are the only ones that, among the subjects within the network, obtain high levels of Eigenvector centrality. The physical meeting places, however, remain secondary to the virtual places. These minor events, where there is less dispersion among the participants, are also those that allow lesser-known authors to relate with the already well-known. The information should be investigated since it may be the result of a distortion in data collection: many artists, demotivated by the difficulty to stay inside the circuit, do not fully collect the information related to their paths, others (the loners) have difficulties in building a solid relational capital and in participating in events requiring expensive and not always paid transfers. Another interesting result is the one obtained from the received reviews that do not seem to be prevalent within the network and for this reason, due to the smaller amount of information collected, they are not considered in this representation. In this case the reason can also be found in the sector's arduousness to be reviewed within non-specialized publications, but also by the sporadic artist skill of retaining information related to the feedback received: only "il Manifesto" is in the list. The role of specialized publishers such as Lieto Colle[11] and Crocetti is still central.

Another way to look at the structure of relations between actors and events is to develop core/periphery measures which allow to identify agents who are strongly connected (the ones in the core) and to assess their role in establishing relations with periphery's partners. Moreover, this approach allows us to strengthen the centrality definition: a node needs to have a high centrality indicator and to be localized in the network's core to be central. It is possible to find that there is not a highly dense core (0.064) in which we can detect a group of poets connected to multiple events and, in turn, surrounded by less connected components. In the core we can find some poets who are heterogeneous: they are used to write using different tools and live in different cities. They are not grouped around a single editor in geographically specific context. Instead, periphery contains disjointed events without common actors and artists who have not worked in the publishing sector and are not co-occur to the same events and, in particular, not attend the most important events (the density of the periphery is 0.011). This structure is more similar to Kadushin artistic circles with "no clear boundaries" and between core and periphery division lines [64].

As we said, poets take part in different events. Participation in festivals, for instance, is an effective way to increase visibility and opportunities to meet publishers, peers, well-established poets, critics, public as well as better define relationships and build new ones [10, 61, 65]. Participation in festivals, readings and slams, often

---

[11] The branch publishing companies, which obtain other levels in all the centrality indices, still play an important role and still represent the natural recipient of the authors' proposals. The high entry costs and the strong selection make them less accessible than online "colleagues" and this lower accessibility also justifies the emerging and central role, but increasingly important, of online magazines. The traditional publishers who are in high positions, according to all the centrality indices, are sectoral and independent except for Mondadori, one of the largest European publishing group (owner of Marcos y Marcos in our selection).

takes place free of charge especially when events are organized by young people. This favors the visibility and the possibility, therefore, of being invited to major events, "to get noticed" from high status player, where the level of promotion and financing allows to a degree the guests to get a salary in addition to a reimbursement. These also represent a fundamental moment of discussion, innovation, and creative growth [62, 65]. The importance of participation in collective events emerges from the interviews: everyone recognizes the fundamental role played by collective events to implement the level of reputational capital and to create new opportunities. All respondents participate in public events.

Observing the matrix that connects poets with collective events (two-mode) it is possible to isolate poets and events that take on greater importance than others and play a strategic role within the network. In the right side of Fig. 2 there are some isolates. These artists have different characteristics and behaviors: Isabella Bordoni, poet and performer from Rimini (IB), and Alberto Pellegatta[12] (AP2) that work abroad are really active but within their own networks. The others (FM, LP, DB, MG2) claim to have great difficulty in moving (for work commitments, logistical problems) and difficulties in relating in public contexts. Inside the main graph, some poets occupy strategic positions and are extremely active: Lidia Riviello (LR) with 17 events, Azzurra d'Agostino (AdA) with 10, Francesca Matteoni (FM2) and Matteo Zattoni (MZ) with 8, Alessandro Raveggi (AR) with 7. These poets reach important positions in terms of centrality.[13]

Despite the number of collective events (in Italy there are a large number of festivals and cultural events, especially during the summer, sometimes with regional character, little cultural impact and organized mainly thanks to local public funding, for entertainment rather than cultural in-depth analysis [16, 66]), most poets participate in a few events of national importance: Parco Poesia (PaPo) with 6 poets, Festival di Armunia (ArMu), Pordenone Legge (PoLeg), Premio Delfini

---

[12]Author of monographs and the many collective works as an editor for the literary and art magazines. He coordinates and directs the EDB magazine.

[13]Lidia Rivello (LR): Degree $(D) = 0.105$ and Eigenvector $(E) = 0.319$ (the second for both indices within the network), is author and host of radio and TV programs (Radiotre, Radiodue, La7, Sky). She collaborates with newspapers, magazines and blogs, curates and organizes events and festivals of poetry and art in Italy and abroad since 1994, collaborating with the magazine "Italian Poetry." She has numerous monographs and is present in numerous Italian and foreign magazines and anthologies. She has worked as a consultant for the cultural programs of the Mibact (Ministry of cultural heritage and tourism) and collaborates with "La Compagnia della Fortezza" in Volterra. She is currently President of the Poetitaly Cultural Associacion. Azzurra d'Agostino (AdA): $D = 0.101$ (3°) and $E = 0.249$ (4°) is a playwright and event organizer, presents in collections of poetry and has published books for children. Matteo Zattoni (MZ): $D = 0.111$ (1°) and $E = 0.199$ (10°), PhD in Sociology of Law, is a researcher at CIRSFID in Bologna, he collaborates with Atelier Magazine. Alessandro Raveggi (AR: $D = 0.084$ (4°) and $E = 0.336$ (1°), novels, short stories, essays and playwright's author, he directs the magazine FLR and is curator for LiberAria Editrice. Francesca Matteoni (FM2): $D = 0.061$ (6°) and $E = 0.258$ (3°), has carried out university research, present in collections and has published at the main publishers of the sector.

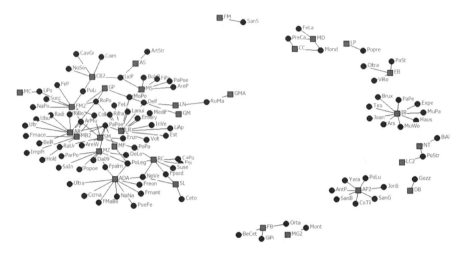

Fig. 2 Lives' affiliation network (squares represent poets and circles represent events)

(Delf), Roma Poesia Festival (RoPo) with 4. These events reach important positions in terms of centrality[14] but not strategic within the entire network.

We see that poets with a relevant position within the network participate in a large number of events. As we said, participation in events has a dual role: inviting central poets within the network allows us to add prestige to the event in itself and make it fit within the author's network relationships, on the other hand the invited author can improve its own economic capital and prestige. The lack of participation (most of the poets participate in a single event), confirmed by the interviews, shows the difficulty in supporting the traditional steps of institutionalization. Although there is a return to orality and direct interaction that is still considered really important today perhaps these are no longer the places appointed (according to the literary canons of the 900) to promote the dissemination of poetry.

Towards the young authors, the successful ones often play as mentors, holding the role of gatekeepers, traditionally occupied by publishers. Writers' versatility has a positive effect in terms of attention from the critics. Receiving journalistic critiques in newspapers or magazines or reviews in specialist magazines is fundamental because it allows the author to be evaluated. Nevertheless, the poets, too focused on their marginality in the contemporary art scene, rarely keep an accurate review of their work, regardless of the nature of comments received. This is because the information received in the specific area is fragmented. Moreover, the publications and critiques received amount are not the only exogenous elements that build the

---

[14]Parco Poesia (PaPo): $D = 0.091$ (16°), $E = 0.149$ (10°); Roma Poesia Festival (RoPo): $D = 0.61$ (27°), $E = 0.138$ (12°); Festival di Armunia (ArMu): $D = 0.061$ (27°), $E = 0.092$ (21°); Pordenone Legge (PoLeg): $D = 0.061$ (27°), $E = 0.082$ (24°), Premio Delfini (Delf): $D = 0.061$ (27°), $E = 0.77$ (26°).

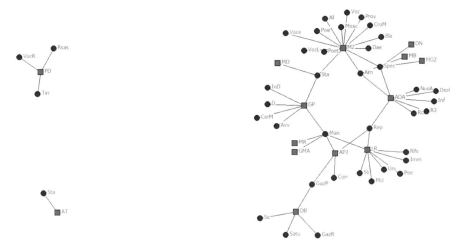

Fig. 3 Reviews' affiliation network (squares represent poets and circles represent events)

attention's level for a new publication. Concerning the relation between artists and critiques received shows a smaller number in the nodes and an ever lower density (Fig. 3). Interesting positions are the one of La Stampa (Sta), il Manifesto (Man), Almanacco dello Specchio di Mondadori (Alm), Lo Specchio della Stampa (Spec), and Repubblica (Rep),[15] newspapers with nationwide circulation focused for the arts representing bridges but not having a strategic position within the network.

Among the authors, only GP,[16] MZ, AdA, and LR have completely collected the reviews received. They are in a strategic position within the network. From this we understand that artists who are established [10] begin to take interest in their position within the field, knowing that critics understand their work and try to implement their prestige's level: they perceive themselves as less marginal within the niche and begin to open up to the public.

While the exercise of poetry appears to be a prestigious activity, the market is limited [66, 67] and given the distribution and marketing costs, the great publishers are often reluctant to invest in contemporary poetry, who do not often realize great sales in the short period. While the poetic practice follows esthetic logics, it is important for the publishing industry to follow economic and market logics [2, 66]. The mainstream publishers limit themselves to publishing well-known artists, with

---

[15]La Stampa (Sta): $D = 0.045$ (36°), $E = 0.091$ (35°); il Manifesto (Man): $D = 0.076$ (20°), $E = 0.117$ (16°); Almanacco dello Specchi di Mondadori (Alm): $D = 0.045$ (36°), 0.063 (35°), Lo Specchio della Stampa (Spec): $D = 0.061$ (27°), $E = 0.074$ (24°), and Repubblica (Rep): $D = 0.045$ (36°), $E = 0.091$ (23°).

[16]Gilda Policastro (GP), italianist and literary critic, is editor for "Allegory Magazine" and has collaborated for various newspapers' cultural pages, from the "Manifesto" to the "Corriere della Sera," as well as lit-blogs and literary information sites, such as "Le parole e le cose" and "Bookdetector." $D = 0.054$ (11°), $E = 0.234$ (6°).

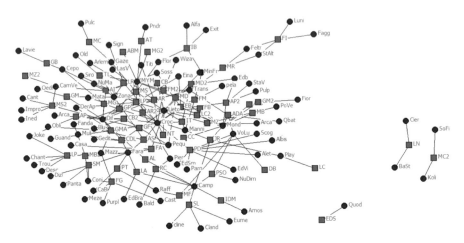

Fig. 4 Publishers' affiliation network (squares represent poets and circles represent events)

their own market because these allow to achieve profits in the long run and leave the task of discovering new talents to minor or independent publishers, who have lower fixed costs in the small runs and limited market distributions [10, 68].

The poetry sector, which represents a very small percentage of the literary one, is characterized by the presence of a large number of highly specialized small publishers and a small number of big general publishers with a section dedicated to historically or well-established poets. From our database 27 poets have published more than 4 books.[17] As for publisher, 11/92 have published more than five authors including in the database.[18] These 11 are also among the 20 subjects that have higher levels of Degree and Eigenvector within the network. Among the authors who have a D greater than 0.054 and an E greater than 0.211, 7 have published at least 3 works with these publishers, as well as having worked with minor publishers. In the most peripheral areas of the network a high number of authors (LP, FG, SM, SL, IDM, FI, IB, AT, MC, MS2) have a similar structure: they are bridge connected with extremely small publishers who have worked only with them, they are the only step to link these publishers to the main network. The meaning of these positions is clear and confirmed by the interviews: many authors access to the publishing market only through self-productions; the most prolific and central authors stay close to the specialized publishers (Fig. 4), which are the most important proponents in this micro-market, endowed with symbolic and economic capital.

The online promotional activity is another important communication channel directly involved with a sensitive to the poetic work people circuit [67]. In this

---

[17]In particular, LP e PD published 8 books, AR, AI, GMA 6, AP2, AS, AR2, CDL, CB2, FM, FG, LP, MS, MD, MS2 5, ADA, EB, FI, FM2, GP, IB, IT, JR, LR, MF, MZ, SM, SL 4.

[18]Lieto Colle: 16 poets, D'if: 14, Crocetti: 13, Mondadori: 12, Marcos y Marcos: 11, Fara and Campanotto: 8, Zona: 7, Transeuropa: 6.

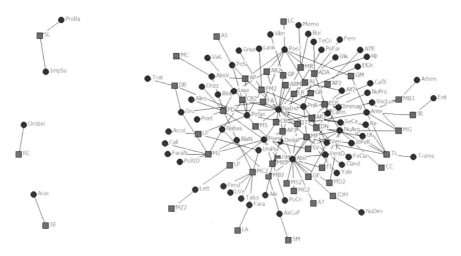

Fig. 5 Online activities' affiliation network (squares represent poets and circles represent events)

case, perhaps thanks to the user-friendly IT tools and to the dissemination of social networks, the authors are particularly active, not only in promoting their work but also in promoting the colleagues one. From this representation some clear positions emerge (Fig. 5): NaInd, Poe2, Atel, Poes, NuArg.[19] These online journals have D greater than 0.106 and E greater than 0.092. These represent the most influential nodes within the network because these are the magazines on which the most central poets have written. The result is not surprising: the online magazine today has a lower entry barrier than traditional publishers and their existence fosters the producers and cultural consumers' expansion at the same time. The writing's greater possibilities, however, lead to a dispersion of energy, which in our network is represented by authors and magazines' multitude that is in the outer part of the network: non-prolific authors and very small magazines (or local) without real impact on the network. Indeed, only 15 authors have published with more than 3 online journals.[20] The dynamic is the same seen on traditional publishers: authors with similar structures are bridges that have contributed to a single magazine that thanks to them is connected to the network (MB3, JR, TL, IDM, MC2, MZ, LP, DB); at the same time, particularly detached from this practice and isolated authors (MC, AS, LC, SM, LA, MZ2) stay connected to the network thanks to their only online contribution, authors having the highest levels of centrality are closer to the magazines with higher levels of centrality (this applies to both D and E) mutually reinforcing their position in the network.

---

[19]Twenty-two poets wrote on Nazione Indiana (NaInd), 16 on Atelier (Atel), 9 on Nuovi Argomenti (NuArg), 8 on Poesia (Poes), 7 su Poesia 2.0 (Poe2).

[20]Among these: AR with 11, MF and PD with 9, AdA with 8, MZ and MC2 with 7, DN and TL with 6.

4.2   A Multivariate Approach for Studying Poetic Production

MCA has been performed in order to understand deeply the relationships between poets and their activities. Figure 6 shows only the first two factors explaining the highest inertia's percentages, that is, the first axis is 20.1% in the first axis and 12.2% in the second. In this space, each actor corresponds to a point and the inter-point distances best approximate the chi-square distances among actor profiles in the original space and represent the actors' relative positions in the network [20]. Therefore, two close points in the map suggest similar characteristics; more specifically, two poets corresponding to two close points reveal similar participation patterns, while if two events are close, they share the same characteristics. Moreover, the proximity in the space of points that correspond to poets and events reveals the preference of those poets towards specific events and therefore specific participation models. Finally, the points (actors and events) that are next to the axes' origin have common characteristics and participation's approach because they are closer to the average actor's profile located in the axes' origin.

In the first overview, the first axis is named "Symbolic Capital" because it is determined by the most contribution of those poets and events that are strictly specialized or in the early stages of their career path. The area on the right includes poets whose "side job" is not in the publishing sector. Poets ADA, MF, GP, MZ and events Lieto Colle, Nazione Indiana, Absolute Poetry, Nuovi Argomenti and Crocetti define the second axis. Along that one the most central activities (even institutionalized) are concentrated. The authors present play "side jobs" within the publishing sector, have a greater number of publications and their writings are heterogeneous (they do not write only poetry). The events are mainly online journals and traditional publishers: live events are completely missing. The second axis was indeed named "Economic Capital" and describes the subjects who are in an advanced state of their career, of "social aging" [30].

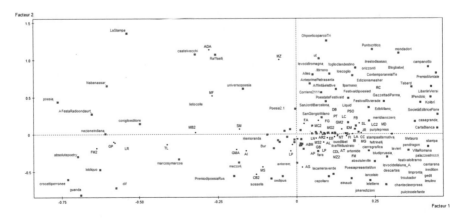

Fig. 6 Events and poets' MCA map (squares represent events and circles represent poets)

Specifically, we can distinguish two well-separated actor subgroups. In the first group those actors and events that are on the left side of the space, while in the second group the right side. In the lower left part of the chart there are big publishers who have a sector dedicated exclusively to poetry, including Crocetti, Perrone, Bur, Marcos y Marcos, Guanda, and online magazines, more precisely Nazione Indiana, Universopoesia, Nabanassar, and Absolute Poetry that could be defined as a hub, an autonomous subfield within the field of literature [69]. Next to these central events, poets already have a career going formed (FM2, GP, LR, TL), which a richer curriculum that counts a greater number of publications and a significant number of reviews. In the upper left area, instead, we find niche publishers (Sossella, Memoranda, Mazzoli, Raffaelli, Castelvecchi, Coniglio Editore, Lietocolle, Oedipus, d'if) and La Stampa, newspaper with a large section dedicated to the arts in which the poets received the highest number of reviews. In this case, we can observe how some poets (MF, MB2, SM, and ADA) tend to concentrate their poetic production through more traditional channels. The subjects that are on this left side are established and institutionalized; their level of relational capital is high as much as their symbolic and economic capital. The right side is distinguished by a minor characterization: poets with less rich curriculum and events to which poets have participated with less frequency. The placing of most events and in particular of the poets near the origin of the axes is interesting: it indicates the participation in the same set of events indiscriminately, therefore, a common participation's model, less defined than that one described in the left side of the chart. It can be said that the left side is the place where things are done today, with characters who carry out many activities in the sector, strongly relational and embedded to an institutional level, but also generalists: they write poetry, novels, dramas. This does not affect their writing's quality but rather places them within the wider literary sector. The behavioral pattern, which this representation describes, is clear: most poets, with a weaker curriculum and who occupied a more peripheral position in the network, share the same approach to the world of poetry, participate in a multiplicity of secondary events, publish in self-produced magazines (both online and on paper); on the other hand, the most relevant subjects within the niche (whose symbolic content remains high) are few, but they are the ones that attract most of the central subjects.

In conclusion, we can say that the multiplicity of tools which can be used to stay present within the poetic field today allows poets to live the trade-off between economic and symbolic capital in a less stringent way: no one is exclusively dedicated to cultural activity, a common destiny today to many intellectual works, but this does not minimize the quality of the contents' proposed.

5   Conclusions

In the art world, artists are often forced to diversify their activities [70] to earn and promote themselves. The poetry's context, more marginal than the publishing's broader one, does not escape from this rule. In this work we have investigated, through the Social Network Analysis' tools, the sub-group of young Italian poets. In addition to the marginality of the production context there is the young people's marginal position, who, as in other sectors of cultural production, struggle to make their voices heard and to see their quality recognized from the establishment. However, the high barriers at the threshold of the publishing market are faced by the young poets in a creative way: as mentioned in this work, the artists are able to develop their activity through self-production, mutual support, creation of new small publishers and, above all, through communication and online work. In that respect, there are spaces not yet occupied by mainstream and spaces in which it is possible to bravely promote activities, in a context of dispersed market and niche not otherwise attainable, except by high promotional costs. In this sense, within traditional economic logics, poets find a logical space for reciprocity, thanks to which it is possible, more easily than through a solitary individual work focused exclusively on improving one's art, to find communication channels and opportunities for artistic expression. The presence of reciprocity and sociability logic, which allows the construction of informal networks capable of sustaining the individuals' careers, characterizes the cultural good poetry as a collective good, since the poetic result does not depend only on the author's isolated genius but from its ability to cooperate and its capability to fall within the relational circuits, to support the work. Therefore, action is also the result of collective support without which poetic work's production and distribution would not be possible: poetry is still a collective and cooperative action [7].

This study confirms how difficult it is for a young poet to fit into a consolidated structure, such as the one characterizing the publishing sector, but at the same time as the marginality of the aforementioned sector makes it more permeable and attentive to the new than the traditional publishing sector, whose production costs no longer allow exposure to the discovery risk.

A multivariate approach for studying actors and events shows us how the most active artists focus on relevant activities: traditional, specialized publishers, and online magazines. Remarking the difference between a larger group of poets, homogeneous and which tends to participate in numerous events scattered throughout the national territory, less important in terms of participations and network, and a group, more restricted, which is characterized by a more specific model of participation, formed mainly by relationship with publishers, online and paper magazines that can offer more opportunities to the more centrals in the network. Emerging circuits that focus on the role of publishers and their effective complement, the online magazines, which give the opportunity to publish (at lower costs) to a greater number of artists and which have replaced the research offices of publishers, represent the place where traditional publishers scout and find talent. The work of cultural collectives and online magazines represents today, with their opening to mass and middle culture productions, the springboard for young artists and the place to go and to look for something new that can, with careful support, be transformed into an artistic or editorial success. This nourishes relational capital as well as the symbolic one. Working on multiple areas within the same sector allows the poet to live in a less stringent way the trade-off between symbolic and economic capital, traditionally illustrated by Bourdieu. This trade-off is outdated in the contemporary context: quality is accompanied by distribution thanks to online tools. Moreover, the barriers' breaking down to entry favored by online tools makes it possible to increase the number of producers and, in a market as pure as the poetry one is, expanding producers means increasing readers and ensuring the survival of the niche.

References

1. Bourdieu, P.: The Field of Cultural Production: Essays on Art and Literature. Columbia University Press, New York (1993)
2. Bourdieu, P.: La Distinction. Minuit, Paris (1979)
3. Collot, M.: Anthologie de la poésie française. Gallimard, Paris (2000)
4. Bourdieu, P., Wacquant, L.: An Invitation to Reflexive Sociology. University of Chicago Press, Chicago (1992)
5. Eco, U.: Apocalittici e Integrati. Bompiani, Milano (1964)
6. Hall, P.M.: Interactionism and the study of social organization. Sociol. Quart. **28**(1), 1–22 (1987)
7. Becker, H.S.: Art Worlds. University of California Press, California (1982)
8. Adorno, T.: The Culture Industry. Routledge, London (1991)
9. Lewis, W.S.: Preface. In: Remer, T.G. (ed.) Serendipity and the Three Princes: From the Peregrinaggio of 1557. University of Oklahoma Press, Oklahoma (1965)
10. Craig, A., Dubois, S.: Between art and money: the social space of public readings in contemporary poetry economies and careers. Poetics. **38**(5), 441–460 (2010)
11. Borgatti, S.P., Everett, M.G., Johnson, J.: Analyzing Social Networks. Sage, London (2013)
12. Scott, J.: Social Network Analysis. A Handbook. Sage, London (2000)
13. Wasserman, S., Faust, K.: Social Network Analysis: Methods and Applications. Cambridge University Press, New York (1994)
14. De Nooy, W.: Fields and networks: correspondence analysis and social network analysis in the framework of field theory. Poetics. **31**, 305–327 (2003)
15. Giuffre, K.A.: Mental maps: social networks and the language of critical reviews. Sociol. Inq. **71**, 381–393 (2001)
16. Pedrini, S., Felaco, C.: Il panorama della giovane poesia italiana: un'analisi di social network. In: Cavallo, M., Marchettini, M., Pedrini, S. (eds.) Lo swing delle città. Clueb, Bologna (2018)
17. Serino, M.: Reti di collaborazione tra teatri di produzione in Campania. Sociol. Lavoro. **138**, 121–137 (2015)
18. Serino, M.: Reti culturali in una prospettiva multidimensionale: Il campo teatrale in Campania. FrancoAngeli, Milano (2018)
19. Serino, M., D'Ambrosio, D., Ragozini, G.: Bridging social network analysis and field theory through multidimensional data analysis: the case of the theatrical field. Poetics. **62**, 66–80 (2017)
20. D'Esposito, M.R., De Stefano, D., Ragozini, G.: On the use of multiple correspondence analysis to visually explore affiliation networks. Soc. Netw. **38**, 28–40 (2014)
21. Benhamou, F.: Economie de la culture. La Découverte, Paris (1996)
22. Istat: La produzione e la lettura di libri in Italia, Roma. https://www.istat.it/it/archivio/225610 (2017)
23. Lahire, B.: La culture des individus, dissonances culturelles et distinctions. La Découverte, Paris (2004)
24. Viala, A.: Naissance de l'écrivain, sociologie de la littérature à l'âge classique. Minuit, Paris (1985)
25. Mazzoni, G.: Sulla Poesia Moderna. Il Mulino, Bologna (2005)

26. Zublena, P.: e la Lingua è Visitata dalla Neve (alter voci). In: Morando, S., Verdino, S. (eds.) La Tua Voce è Plurale. Per Eugenio De Signoribus, pp. 61–64. Interlinea, Novare (2017)
27. Perniola, M.: Contro la Comunicazione. Einaudi, Torino (2004)
28. Giovannetti, P., Lavezzi, G.: La Metrica Italiana Contemporanea. Carocci, Milano (2010)
29. McDonald, D.: Masscult e Midcult. Piano B, Prato (2018)
30. Bourdieu, P.: The Rules of Art: Genesis and Structure of the Literary Field. Stanford University Press, Stanford (1996)
31. Cheal, D.: Gift Economy. Routledge, London (1988)
32. MacIntyre, A.: After Virtue: A Study in Moral Theory, 2nd edn. Duckworth, London (1985)
33. Simmel, G., Hughes, E.C.: The sociology of sociability. Am. J. Sociol. **55**(3), 254–261 (1949)
34. Granovetter, M.: Action economique et structure sociale. Le probleme de l'encastrement. In: Le marche´ autrement, pp. 75–114. Descle'e de Brouwer, Paris (2000)
35. Uzzi, B.: The sources and consequences of embeddedness for the economic performance of organizations: the network effect. Am. Sociol. Rev. **61**(4), 674–698 (1996)
36. Bourdieu, P.: Campo del potere e campo intellettuale. Manifestolibri, Roma (2002)
37. Prigent, C.: A quoi bon encore des poetesï. Editions POL, Paris (1996)
38. Guillaume, D.: Poésies et poétiques contemporaines. Le Temps qu'il fait, Cognac (2002)
39. Anheier, H. K., Gerhards, J.: The acknowledgment of literary influence: A structural analysis of a German literary network. In Sociological Forum, vol. 6(1), pp. 137–156. Kluwer Academic Publishers New York (1991).
40. Emirbayer, M.: Manifesto for a relational sociology. Am. J. Sociol. **103**(2), 281–317 (1997)
41. Bourdieu, P.: Meditazioni pascaliane. Feltrinelli, Milano (1998)
42. Laumann, E.O., Marsden, P.V., Prensky, D.: The boundary specification problem in social network analysis. In: Burt, R.S., Minor, M.J. (eds.) Applied Network Analysis: A Methodological Introduction. Sage, Beverly Hills (1989)
43. Antoci, F., Sabatini, F., Sodini, M.: See you on Facebook! A framework for analyzing the role of computer-mediated interaction in the evolution of social capital. J. Socio Econ. **41**, 541–547 (2012)
44. Borgatti, S.P., Everett, M.G., Freeman, L.C.: UCINET 6 for Windows: Software for Social Network Analysis. Analytic Technologies, Harvard (2002)
45. Borgatti, S.P., Everett, M.G.: Models of core/periphery structures. Soc. Netw. **21**, 375–395 (2000)
46. Frachisse, D., Billand, P., Massard, N.: The sixth framework program as an affiliation network: representation and analysis (March 1, 2008). FEEM Working Paper No. 32.2008. Available at SSRN: https://ssrn.com/abstract=1117966 or https://doi.org/10.2139/ssrn.1117966 (2008).
47. Bonacich, P.: Simultaneous group and individual centralities. Soc. Netw. **13**, 155–168 (1991)
48. Faust, K.: Centrality in affiliation networks. Soc. Netw. **19**(2), 157–191 (1997)
49. Freeman, L.C.: Centrality in social networks: I. Conceptual clarification. Soc. Netw. **2**, 215–239 (1979)
50. Fernandez, R.M., McAdam, D.: Social networks and social movements: multiorganizational fields and recruitment to Mississippi freedom summer. Sociol. Forum. **3**, 357–382 (1988)
51. Mariolis, P., Jones, M.H.: Centrality in corporate interlock networks: Reliability and stability. Admin. Sci. Quart. **27**, 571–584 (1982)
52. Borgatti, S.P., Everett, M.G.: Network analysis of 2-mode data. Soc. Netw. **19**, 243–269 (1997)
53. Dramalidis, A., Markos, A.: Subset multiple correspondence analysis as a tool for visualizing affiliation networks. J. Anal. Inf. Process. **4**(2), 81–89 (2016)
54. Roberts, J.M.: Correspondence analysis of two-mode network data. Soc. Netw. **22**, 65–72 (2000)
55. Benzécri, J.P.: L'Analyse des Données, Tome 2: L'Analyse des Correspondences. Dunod, Paris (1973)
56. Greenacre, M.: Theory and Applications of Correspondence Analysis. Academic Press, New York (1984)
57. Lebart, L., Morineau, A., Piron, M.: Statistique Esploratoire Multidimensionnelle. Dunod, Paris (1995)

58. Ragozini, G., De Stefano, D., D'esposito, M.R.: Multiple factor analysis for time-varying two-mode networks. Netw. Sci. **3**(1), 18–36 (2015)
59. Borgatti, S.P., Halgin, D.: Analyzing affiliation networks. In: Carrington, P., Scott, J. (eds.) The Sage Handbook of Social Network Analysis, pp. 417–433. SAGE Publications Ltd, London (2011)
60. Faust, K.: Using correspondence analysis for joint displays of affiliation networks. In: Carrington, P.J., Scott, J., Wasserman, S. (eds.) Models and Methods in Social Network Analysis, pp. 117–147. Cambridge University Press, Cambridge (2005)
61. Gregory, H.: The quiet revolution of the poetry slam: the sustainability of cultural capital in the light of changing artistic conventions. Ethnogr. Educ. **3**(1), 63–80 (2008)
62. Farrell, M.P.: Collaborative Circles: Friendship Dynamics and Creative Work. University of Chicago Press, Chicago (2001)
63. Stolarick, K., Florida, R.: Creativity, connections and innovation: a study of linkages in the Montréal region. Environ. Plan. A. **38**(10), 1799–1817 (2006)
64. Kadushin, C.: Networks and circles in the production of culture. Am. Behav. Sci. **19**(6), 769–784 (1976)
65. Bronwyn, L.: Trends in poetry publishing: 1995–2008 [Paper delivered at the Australian Poetry Festival, 6th: 2008: Sydney] (2007).
66. Pedrini, S.: Il pop rock bolognese tra il 1978 e il 1992: un'analisi di network per descrivere il milieu culturale e creativo della città di Bologna. In: Cavallo, M., Marchettini, M., Pedrini, S. (eds.) Lo swing delle città, pp. 161–190. Clueb, Bologna (2018)
67. Jarrety, M.: La poe'sie française du Moyen-Age au XXe sie'cle. PUF, Paris (2007)
68. Dubois, S.: The French poetry economy. Int. J. Arts Manag. **9**(1), 23–34 (2006)
69. Bortolotti, G.: Blog e letteratura, Nazione Indiana. https://www.nazioneindiana.com/ (2008).
70. Menger, P.M.: Artistic labor markets and careers. Ann. Rev. Sociol. **25**(1), 541–574 (1999)

Multilayer Network Analysis
of Innovation Intermediaries' Activities

Margherita Russo, Annalisa Caloffi, Riccardo Righi, Simone Righi, and Federica Rossi

Abstract Policymakers wishing to enhance innovation processes in small and medium-sized enterprises increasingly channel their interventions through innovation intermediaries. However, limited empirical research exists regarding the activities and performance of intermediaries, with most contributions taking a qualitative approach and focusing on the role of intermediaries as brokers. In this paper, we analyse the extent to which innovation intermediaries, through their engagement in different activities, support the creation of communities of other agents. We use multilayer network analysis techniques to simultaneously represent the many types of interactions promoted by intermediaries. Furthermore, by originally applying the Infomap algorithm to our multilayer network, we assess the contribution of the agents involved in different activities promoted by intermediaries, and we identify the emerging multilayer communities and the intercohesive agents that span across several communities. Our analysis highlights the potential and the critical features of multilayer analysis for policy design and evaluation.

The Authors are members of CAPP-Centro di Analisi delle Politiche Pubbliche, Department of Economics, University of Modena and Reggio Emilia, Italy.

M. Russo (✉)
Department of Economics, University of Modena and Reggio Emilia, Modena, Italy
e-mail: margherita.russo@unimore.it

A. Caloffi
Department of Economics and Management, University of Florence, Florence, Italy

R. Righi
European Commission, Joint Research Centre (JRC), Seville, Spain

S. Righi
Department of Computer Science, University College London, London, UK

"Lendület" Research Center for Education and Network Studies (RECENS),
Hungarian Academy of Sciences, Budapest, Hungary

F. Rossi
Birkbeck, University of London, London, UK

© Springer Nature Switzerland AG 2020
G. Ragozini, M. P. Vitale (eds.), *Challenges in Social Network Research*,
Lecture Notes in Social Networks, https://doi.org/10.1007/978-3-030-31463-7_12

Keywords Multilayer and multiplex networks · Overlapping communities ·
Regional innovation policy · Innovation intermediaries · Regional innovation
system

1 Issue: Intermediaries in Innovation Processes

The term 'innovation intermediary' is used in the innovation literature, particularly
in the systems-of-innovation perspective, to identify a varied set of organisations
whose main mission is to support innovation processes involving other organisations. They are generally needed in the context of rapid changes or when small
firms are not able to keep the pace of change in their technical and economic
environment. Intermediary functions are performed by private organisations [1],
such as innovation consultants [2], innovation brokers [2, 3], and knowledge-
intensive business service providers [4], as well as by publicly or public-privately
funded organisations such as regional institutions [5], research-industry liaison
offices and science parks [6]. Examples of innovation intermediaries funded with
substantial public contributions are the Competitiveness Poles in France, the
Innovation Networks in Denmark, the Strategic Centres for Science, Technology
and Innovation in Finland, the Technology Catapults in the UK, and the Innovation
Poles in Italy [7].

Early academic studies of innovation intermediaries emphasised brokering as
their central concern.[1] Subsequent literature has suggested that brokers should be
considered 'facilitators of innovation', as they support innovation processes, but
the innovations neither originate from nor are transferred by them [8, 9]. Other
types of intermediaries act as either 'sources of innovation' (i.e. they play a major
role in initiating and developing an innovation) or as 'carriers of innovation'
(i.e. they transfer an innovation that does not originate from them) [8]. Recent
studies move further away from a view of intermediaries as simple providers of
connections between other agents positioned at different stages of linear innovation
processes, but see intermediaries as enhancing the innovation capacity of many
actors embedded within complex systems of interactions. For example, [10] defines
innovation intermediaries as 'organisations or groups within organisations that work
to enable innovation, either directly by enabling the innovativeness of one or more
firms, or indirectly by enhancing the innovative capacity of regions, nations, or
sectors'. In the context of their complex role, intermediaries engage in a plurality
of activities that can include, among others, technology foresight and technology
scouting, supplier selection, research and development (R&D) partnership formation, technical assistance in the realisation of R&D projects, dissemination and
commercialisation of results, and technology transfer. Through all of these activities,

---

[1]For example, Howells [4] proposed a functional definition of innovation intermediary as '[a]n
organisation or body that acts [as] an agent or broker in any aspect of the innovation process
between two or more parties'.

even those that are not specifically aimed at networking, intermediaries engage
with a wide variety of other organisations, and contribute to the formation of
communities of organisations that share a common orientation towards open and
collaborative innovation. Despite the growing interest in innovation intermediaries,
to the best of our knowledge no studies have yet analysed the extent to which
intermediaries support the creation of communities of other agents, and the extent
to which they do so through the different activities they perform. Thanks to original
data relating to a set of publicly funded innovation intermediaries in the region of
Tuscany (Italy), we reconstruct the multilayer network of interactions generated
by the set of 12 regional innovation intermediaries through a range of different
activities that include, among others, the provision of knowledge-intensive services,
the setup of collaborative projects, and the promotion of events for members. We
then apply the Infomap algorithm [11–13] to detect the emergence of communities
around those intermediaries and the actors that play the most central roles within
those communities. The composition of communities, by type of agent and by type
of activity, and the presence of agents belonging to more than one community are
of utmost interest in designing innovation policies that aim to support regional
innovation systems. In particular, we explore the following questions:

– *Role of agents in information dynamics*: to what extent were the innovation
  intermediaries the most relevant agents in setting up a range of different
  interactions, and who were, if any, other pivotal agents?
– *Structure of communities*: What was the structure of the overlapping communities
  of agents mobilised by the innovation intermediaries?

Our analysis is original with respect to the existing literature, in two main ways.
First, we develop an original empirical analysis of intermediaries' activities, using
multilayer network representations. Second, we present a new application of the
Infomap algorithm for multilayer networks. Such an algorithm allows us to address
our two research questions as it provides information regarding (1) the role of
each node in the considered network, (2) the detection of communities of agents.
Moreover, the use of the Infomap algorithm is particularly appropriate in our
context, as information flows are among the elements that determine and influence
the formation and the dynamics of socio-economic complex systems, as the one
considered in this work [14]. To answer our research questions, this chapter presents
the empirical dataset, in Sect. 2, and briefly introduces the multilayer methodology
adopted in our analysis, in Sect. 3; Sect. 4 summarises the results with regard to both
the centrality of agents in the different networks and the characteristics of emerging
communities. Some comments on the Infomap methodology, the policy implications
of this approach, and current developments of our research are discussed in Sect. 5.

2 Data: Tuscany Policy Programme 2011–2014

2.1 The Regional Policy

Our analysis focuses on a unique database collecting data on innovation intermediaries, funded through a regional policy intervention, in the Italian region of Tuscany.[2] In the programming period 2007–2013, the regional government of Tuscany funded twelve 'innovation poles'. Each pole was managed by a managing consortium that included several organisations operating in the field of innovation, such as universities, service centres, and enterprises. The managing consortium was led by a leading organisation. In one case only the managing consortium coincided with its leading organisation. Consortium participants would share the use of their research laboratories and facilities and second some of their staff to the poles so that they could carry out specific activities. Poles engaged in the provision of innovation advisory services [15], support to networking or to R&D activities, and other activities that are typically performed by intermediaries. The poles received public funding to provide these services for free to firms in the region. The regional firms, in order to take advantage of these services, had to become a member of the poles. Poles, which were active in the period July 2011–June 2014, were specialised in different technological areas and applications, as listed in Table 1.

The 3181 members associated to poles were mainly manufacturing enterprises (66.3%). The remaining share included companies in the service sector: from traditional services (21.9%) to knowledge-intensive business services (11.8%). Less than one per cent of the members were other types of organisations (e.g. business

Table 1 The 12 innovation poles: key technologies and applications, consortium participants, and member companies, by type

id Innovation pole (consortium's acronym)	Key technologies or applications	Organisations managing the consortia — Types (underline: the type including leader) N (including leader)	University	Public reserch center	Private reserch center	Enterprise	Tech. Park	Service Center	Members of the consortia N. (in 2014)	Types of innovation poles' members ■ Enterprises ▨ Service companies ▫ KIBS 0 100 200 300 400 500 600 700
12 POLITER	ICT Technologies, Telecommunic.& Robotics	13	4	6				<u>3</u>	697	
7 POLIS	Technologies for sustainable cities	8	2	2				<u>4</u>	643	
3 OTIR 2020	Fashion (textiles, apparel, leather, shoes, jewelry)	7			<u>2</u>			5	501	
11 POLO12	Mechanics, particularly for automotive and transport	6			1			<u>5</u>	390	
6 PENTA	Shipbuilding and maritime technology	5					1	<u>4</u>	352	
9 CENTO	Furniture and interior design	6	1					<u>5</u>	322	
10 PIERRE	Renewable energies and energy saving technology	13	5	2	1			<u>5</u>	368	
2 INNOPAPER	Paper	1						<u>1</u>	139	
4 VITA	Life science	8	5	1		<u>1</u>		1	158	
8 NANOXM	Nanotechnologies	6	2	2				<u>2</u>	128	
5 PIETRE	Marble	4	1				1	<u>2</u>	122	
1 OPTOSCANA	Optoelectronics for manuf. & aerospace	2		<u>1</u>				1	92	

[2]The database is available at the following https://doi.org/10.25431/11380_1182469.

associations). The policy was clearly inspired by the Triple Helix approach to innovation, with poles aiming to facilitate the interaction between the three main actors of the innovation system: research, government, enterprises [16]. Moreover, the policymaker was also hoping to promote the development of a regional innovation system [17] that could encourage technology transfer and stimulate the innovation capabilities of regional small and medium-sized enterprises (SMEs).[3]

2.2 The Interactions Promoted by the Poles

The innovation poles performed different kinds of activities, which generated interactions among the poles themselves, the poles' leading organisations, and the members of the poles. By monitoring the development of the poles' activities, we have identified the following types of interactions:

- *Leadership/management.* The interactions between a pole, its managing consortium, and its leading organisation. Note that the same agent (e.g. the same university) can participate in the managing consortia of more than one pole.
- *Membership.* The interactions between a pole and its members. Note that the same regional actor may become a member of more than one pole, in order to get access to the services provided by the managing consortia of each of them.
- *Seconding of staff.* The interactions between the staff of the managing consortium (or leading organisation) who are seconded to the pole, as well as those between such workers and their employers.
- *Service provision.* The interactions between the pole's managing consortium (or leading organisation) and the pole members that benefit from their services.
- *Collaboration agreements.* The interactions between the poles and the organisations that sign the collaboration agreement.
- *Shareholding.* The interactions between the pole's managing consortium (or leading organisation) and their shareholders (i.e. those who legally own one or more shares in the organisations managing the consortium).

2.3 Structuring the Data as a Multilayer Network

Thanks to the identification of these types of interaction, six layers of a single multilayer network (MLN) are generated. Descriptive statistics for each layer and for the aggregate network are presented in Table 2. Out of the 3986 agents involved overall, we have that: 3305 of them are active in just one layer (for the largest part

---

[3]This information was collected during two interviews (on April 2011 and June 2013) with policymakers who designed and managed the policy.

Table 2 Network descriptive statistics, per layer and in the full network

Multilayer Ntw.	Leadership	Membership	Staff	Services	Collaboration	Shareholding	Full Ntw.
# Nodes	63	3181	477	601	166	338	3986
# Edges	193	4465	1147	644	338	358	6113
Density	0.0494	0.0004	0.0051	0.0018	0.0123	0.0031	0.0004
Mean Degree	6.127	2.807	4.809	2.143	4.072	2.118	3.067
# Conn. Components	41	2779	36	30	138	25	2779
Size Giant Components	1	1	1	0.960	1	0.962	1

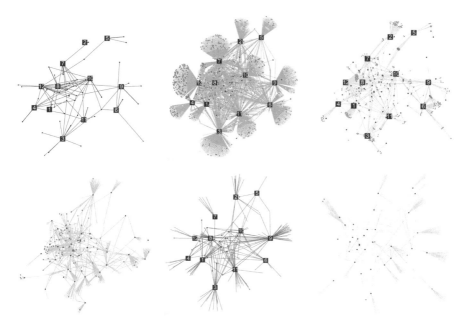

Fig. 1 Graph of 3986 agents, by mode of interaction (layer). *Nodes* have the same coordinates in all graphs. *Nodes' colours*: in all graphs, the nodes representing the poles are shown as black squares, with white figures corresponding to 'id' column in Table 1; the nodes representing the leading organisations/managing consortia are black; the nodes representing workers/consultants are blue; the nodes representing KIBS are orange; yellow nodes represent all other types of agents. *Top Left Panel*: Leading/managing interactions (leading links in black, managing links in red). *Top Centre Panel*: Membership (linkages in grey). *Top Right Panel*: Seconding workers (seconding a worker links in light blue, seconded worker links in dark blue). *Bottom Left Panel*: Service provision (links in green). *Bottom Centre Panel*: Collaboration agreements (links in pink). *Bottom Right Panel*: Shareholding (links in aquamarine)

they are members of the poles); 611 of them are active in two layers; 20 of them are active in three layers; 24 of them are active in four layers; 13 of them are active in five layers; 13 of them are active in six layers.

Figure 1 displays the layers. Note that in this visualisation, each node maintains the same coordinates in all the layers in which it is present. Even if the detail of the single nodes is not visible, Fig. 1 gives an overall view from which it is possible to understand that the different layers are characterised by a different relational structure, since each node always maintains the same position in the various layers.

As we are interested in the analysis of the overall activity of intermediaries, we consider all the six layers. However, instead of considering all interactions indistinctly, i.e. just as connections between agents, the MLN allows us to develop an analysis of the distinct dimensions of interactions.

3 Methodology

3.1 Flows of Information in a Multilayer Network Perspective

Rosvall and Bergstrom [11, 12] introduced a method, based on information theory, to detect communities in complex networks by minimising the length of a two-level description of a simulated flow circulating through the network.[4] Such a method is implemented in an algorithm called 'Infomap' [18]. Infomap solves the main problems with Newman and Girvan [19] in identifying communities of very different sizes and, in addition, it allows the detection of overlapping communities [20, 21], so that each agent/node can belong to more than one of them. Recently, the Infomap algorithm has been extended by De Domenico et al. [13], so as to run over a multilayer—or multiplex—structure[5] by

1. considering in each layer just a specific type of relationships (and the agents/nodes that are involved in these), so as to figure out intra-layer connections;
2. connecting all the projections (i.e. its state-nodes) of the same agent/node (physical-nodes) with inter-layer connections.

Thanks to this implementation, De Domenico et al. [13] show that some features regarding interactions over different dimensions, represented by the layers, can be better investigated and finally unveiled. In addition, as a result of this development, the Infomap multilayer algorithm is set to compute the amount of information flow associated to each state-node, so as to assess the amount of information that agent/nodes collect in each layer in which they have at least one connection. While in network analysis the centrality measures that are most commonly taken into account are betweenness and closeness, in this work we consider the Infomap flow to evaluate agents/nodes centrality in a multilayer structure. As discussed by Solé-Ribalta and De Domenico [23, p. 76], typical centrality measures, e.g. closeness and betweenness, when implemented in a multilayer network allow for a more correct estimation.[6] However, only the more sophisticated structure designed by

---

[4]The first level of description concerns the nodes in which the flow moves, and the second level of description concerns sub-areas of the network, i.e. communities, in which the flow tends to circulate for a long period before exiting. Therefore each detected community maximises the probability of the considered random walk to remain within its boundaries before moving into another community.

[5]As reminded by Arenas and De Domenico [22], historically, the term multiplex was coined to indicate the presence of more than one relationship between the same actors of a social network. The terms 'multiplex' and 'multilayer' are used almost indistinctly as they fundamentally refer to the same concept.

[6]Sole-Ribalta and De Domenico [23, p. 76] discuss the problem of overestimation of closeness centrality in an aggregated network, i.e. a network originally formed by several layers that are all aggregated into a single one.

De Domenico et al. [13] makes it possible to disentangle the impact that each layer has in determining the relevance of each agent/node. Based on De Domenico et al.[13], our analysis describes agents/nodes' role by means of the Infomap flow index. Such an index is decomposable not only with regard to the communities to which the agent/node belongs, but especially with regard to the set of layers in which each agent/node is active. Then, we consider the structure of communities generated by nodes' overlaps (i.e. by the presence of agents/nodes belonging to more than one community), which are detected in the MLN, i.e. mesoscale organisations of agents/nodes that can be both bounded in a single layer or that can extend over several layers. In performing such analysis, we investigate: (1) how each node performs in the process of managing information, with respect to the multiple dimensions of interactions in which it is involved (represented by the layers), and (2) the structure of communities emerging through the presence of agents/nodes belonging to more than one of them. Agents that belong to more than one community have been defined as 'intercohesive agents' [24, 25]: these agents are able to create bridges among communities, so they contribute in the sharing of information and competences between them, and they play a determinant role in the evolution of a community structure.

3.2 Settings of the Algorithm

We analyse the intermediaries' interactions with a focus on multidimensional links. We model the six types of interactions described above, in Sect. 2 and in Fig. 1, as layers of a MLN. Descriptive statistics of agents' activities in the layers are summarised in Table 2. Agents/nodes are present on a layer if they are involved in at least one interaction of the kind corresponding to the layer. In addition, we also consider inter-layers linkages to connect all the projections (in different layers) of the same agent. All connections (both intra and inter-layers) are undirected and unweighted. The algorithm is asked to identify overlapping communities, thus allowing the nodes to belong to more than one of them. This is necessary for our goals, as it makes it possible to establish the presence of connections among communities (through the overlaps) and, in turn, to investigate the whole structure of communities. Finally, the probability with which the random walk simulated by Infomap can jump, moving from one point of the MLN to another random point of the MLN, is set equal to 0.15, as in the PageRank algorithm[7] [26].

---

[7]This is also the value that in [11, 12] is used as default value to run and test Infomap.

4 Main Results

The multilayer Infomap algorithm allows the assessment of the contribution of each layer and each agent, or groups of agents, to the generation of the total Infomap flow. From Table 3, we observe that the six layers have different importance in generating the information flow: 60% is generated by the membership to the poles; almost 16% is due to the network of interactions across poles through workers and consultants; 10.6% is the share activated by the provision of services from managing organisations to the poles' members; almost 6% of the Infomap flow reinforces the connections across poles through their indirect links due to the many organisations (mainly local government or public institutions, like Chambers of Commerce) owning shares in the organisations that manage the poles.

Let us now consider the Infomap flow generated by the innovation poles: almost 37% of the total multilayer flow is generated by the interactions in which poles are engaged, and the largest part of it (almost 30%) is created through poles' membership. With regard to the 46 organisations leading and managing the poles, we observe that their involvement in the information dynamics, as measured by the Infomap multilayer flow, is due not only to the provision of services to member companies (5.82%) and workers to the poles (4.49%), but also to the many connections among their shareholders (3.53%).

When the multilayer analysis is run, it produces 71 communities, 63 of which are overlapping (Fig. 4), with 605 intercohesive agents (15% of the total) mainly active in two communities. Table 4 lists the types of intercohesive agents, showing that only poles belong to three communities (and only in three cases), while the majority of agents belonging to two communities are manufacturing companies (424 firms) and knowledge-intensive business services (KIBS) (98 firms): manufacturing companies are simply members of two poles, while KIBS are active in demanding services.

The multilayer analysis is informative with regard to the structural aspects of the network: it disentangles communities presenting a higher probability of having

Table 3 Percentage of Infomap flow, by layer and group of agents: poles (first column), leading organisations and 46 organisations belonging to the managing consortia (second column), other agents (third column)

Layers	Types of agent			
	12 Poles	Managing consortia	All other organisations	All agents
Leadership	0.93	1.83	0.04	2.8
Membership	29.66	1.28	29.30	60.2
Staff	4.16	4.49	7.21	15.9
Services	0.00	5.82	4.75	10.6
Collaboration	1.84	1.08	1.50	4.4
Shareholding	0.00	3.53	2.57	6.1
Total	36.59	18.03	45.38	100.0

Table 4 Number of nodes, by type, belonging to one, or two, or three communities resulting from the implementation of Infomap over the MLN

	# of communities			
	1	2	3	Total
Poles	3	6	3	12
Territorial pubic bodies	112	4	0	116
Chamber of commerce	8	1	0	9
University	78	4	0	82
Public research institutions	20	1	0	21
Private research institutions	19	0	0	19
Services centre	19	9	0	28
Manufacturing company	1991	424	0	2415
Service company	194	46	0	240
KIBS	406	98	0	504
Company association	67	8	0	75
Other	44	1	0	45
Workers/consultants	420	0	0	420
Total	3381	602	3	3986

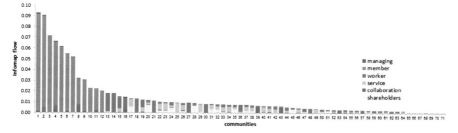

Fig. 2 Multilayer Infomap flow associated to the resulting 71 communities. Colours highlight information regarding the amount of flow generated by each of the six layers. Communities are ordered in decreasing order of total flow

specific connections. Figure 2 shows the relative importance of the multilayer communities and highlights the type of relationships that contributed to their emergence. Figure 3 highlights the composition of communities by type of agent: the first twelve communities (with 3202 agents) account for 62% of total Infomap flow and have links with the other 60 communities. Each of them has one pole as a pivotal agent. In these communities, the largest share of agents are active only in the layer of membership, the one of the poles they belong to. The analysis shows that most of the agents involved have not exploited the potential of joining the innovation poles. Free membership was not a sufficient incentive to stimulate poles' members demand for services (which was the goal of innovation policy). The organisations that managed the poles and offered the services were only able to exploit part of the demand for services that the poles' members generated. This result, examined in detail by Caloffi et al. [27], opens up the issue of the most appropriate incentives to orient the behaviour of the target beneficiaries of the policy.

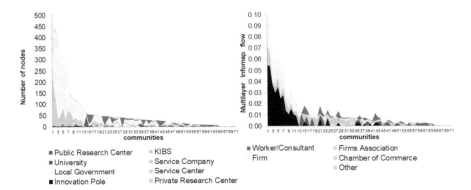

Fig. 3 *Left Panel*: The 71 multilayer communities: number of agents/nodes by type. *Right panel*: Multilayer Infomap flow associated to the resulting communities, with information regarding the amount of flow generated by type of agent. Communities are ordered from the largest to the smallest (in terms of number of agents involved)

Overlapping of nodes across communities generates a network structure with 63 communities (out of 71) belonging to the largest component (Fig. 4). The poles largely contribute to the overlapping with other communities, as can be seen from Fig. 4 (where the twelve communities centred on poles are represented only with their label). The other 51 overlapping communities are represented as nodes whose size is proportional to Infomap flow, with each slice representing the share of flow per layer. Edges' width is proportional to the number of agents in common between communities. Results on the overlapping communities provide a way to characterise who are the agents in each community, and the specific activities through which they are connected; a significant result is also the information on which are the most connected communities.

5   Lessons from the Multilayer Analysis

Before concluding with a summary of the main results emerging from the Infomap multilayer analysis, some methodological and analytical issues should be stressed with regard to the application of the Infomap multilayer algorithm. First of all, identification of layers is crucial for effective analysis, and weights of linkages are a critical aspect of this analysis. Let us take the case of services: the algorithm computes for this layer 10.6% of the total multilayer Infomap flow, but when we use all the information in our dataset, distinguishing each type of service provided, we obtain a rather different result. Indeed, Infomap flow is affected by the repetition of an interaction, which is in principle not a problem per se, but we should weight the information flow also for all the other interaction streams under analysis (e.g. by weighting the share of shareholdings, the number of meetings in the board of directors in each consortium, etc.). In our case study, none of these additional

Fig. 4 Network of communities: the largest component encompassing 63 communities. Nodes' size is proportional to Infomap flow. The 12 communities centred on poles' memberships (the largest 12 communities) are not represented in proportion to their size, and they are labelled with the name of the corresponding pole. Slices represent the share of flow by layer (see the legend for the colour by layer). Edges' width is proportional to the number of agents in common between communities

pieces of information would produce a comparable set of weights to be associated to interactions in each layer and this is why we have excluded any reference to weights for all the relationships under analysis. In general, weights of relationships might not be comparable and information might have different granularity. For this reason, careful attention must be paid to the identification of layers in implementing the Infomap algorithm.[8]

With regard to our research questions on intermediaries' activities and the design of innovation policy, three main results are drawn by the specific metrics generated by the Infomap multilayer algorithm, which we could not have singled out through any other measure of centrality or method to identify communities of interacting agents. First, the Infomap algorithm allows us to assess to what extent were the innovation poles supported by the multidimensional activity of their managing consortia (universities, research centres, and service centres) and which of these consortia were more effective in mobilising the regional innovation system through the set of specific activities they performed. The organisations participating in the managing consortia generated interactions both within each of the poles in which they were involved (as part of the managing consortium), and across their other connections with other managing consortia or agents embedded in the regional system of innovation (such as KIBS). Moreover, since shareholding was modelled as one of the layers, the Infomap flow weights also the relative importance of the indirect connections enabled by the policy that were provided through the shareholders of the managing consortia involved in the poles.

Second, by distinguishing the amount of information flow generated by the innovation poles through their different activities, we are able to identify which activities contribute most to the generation of information and the creation of communities. We can see, in particular, that simple membership of the poles, rather than access to services, generates most of the information flow. By further disentangling these activities at a more fine-grained level (for example, by distinguishing different types of services), it might be possible to evaluate which of the different poles' activities have a greater impact on the innovation system.

A third result is the detection of overlapping communities of agents mobilised by the regional policy. The Infomap multilayer analysis provides metrics to disentangle in each of the communities the relative importance of the six types of relationship and of the various agents involved. The identification of the resulting intercohesive agents unveils who are the players of the innovation ecosystem that are supported by the regional policy, and through which channels those agents support its potential development. In the design of an innovation policy aiming at enhancing system transformation, the description of all these features might support the definition of more focused criteria to define the activities to implement and the types of agents to involve. The multilayer analysis of the interactions generated by the innovation intermediary could also be usefully deployed as an evaluation instrument in order to illustrate the breadth of engagement of the intermediary and also to compare the different compositions and the different information flows of the MLNs generated by different intermediaries.

---

[8]Directedness, too, may affect the result, although in the case study it did not affect the ranking of agents according to the Infomap flow (results not reported).

Building on these results, a current development of our research is more focused on the economic implications of the emerging structure of communities mobilised by the innovation policy and it concerns the analysis of the economic performance of the intercohesive agents.

Acknowledgements This paper has been developed in the research project 'Poli.in Analysis and modelling of innovation poles in Tuscany' (www.poliinovazione.unimore.it), co-funded by Tuscany's Regional Administration and University of Modena and Reggio Emilia, Italy. For their comments to preliminary versions of this paper, we wish to thank the participants in the 1st EAEPE—RA[X] Workshop 'New Frontiers and Methodological Advances in Cooperation and Network Research', November 2–3, 2015 in Essen, Germany; the Conference 'Networks, Complexity and Economic Development', organised by the Hungarian Academy of Sciences, Research Centre for Economic and Regional Studies MTA KRTK, 30 November to 1 December 2015, Budapest, Hungary; the ARS'17 Conference, held in Naples, Italy, May 15–17, and the EUSN 2017 Conference, held in Mainz, Germany, September 26–29. Simone Righi acknowledges funding from the European Research Council (ERC) under the European Union's Horizon 2020 research and innovation programme (grant agreement No 648693).

References

1. Smits, R., Kuhlmann, S.: The rise of systemic instruments in innovation policy. Int. J. Foresight Innov. Policy **1**, 4–32 (2004)
2. Klerkx, L., Leeuwis, C.: Establishment and embedding of innovation brokers at different innovation system levels: insights from the Dutch agricultural sector. Technol. Forecast. Soc. Chang. **76**, 849–860 (2009)
3. Batterink, M.H., Wubben, E.F.M., Klerkx, L., Omta, S.: Orchestrating innovation networks: the case of innovation brokers in the agri-food sector. Entrep. Reg. Dev. **22**(1), 47–76 (2010)
4. Howells, J.: Intermediation and the role of intermediaries in innovation. Res. Policy **35**(5), 715–728 (2006)
5. Hargadon, A., Sutton, R.I.: Technology brokering and innovation in a product development firm. Adm. Sci. Q., **42**(4), 716–749 (1997)
6. Hansson, F., Husted, K., Vestergaard, J.: Second generation science parks: from structural holes jockeys to social capital catalysts of the knowledge society. Technovation **25**(9), 1039–1049 (2005)
7. Russo, M., Caloffi, A., Rossi, F., Fiordelmondo, V., Ghinoi, S.: Evaluating the performance of innovation intermediaries: insights from the experience of Tuscany's innovation poles. fteval— J. Res. Technol. Policy Eval. **41**, 15–24 (2015)
8. Den Hertog, P.: Knowledge-intensive business services as co-producers of innovation. Int. J. Innov. Manag. **4**(4), 491–528 (2000)
9. Van Lente, H., Hekkert, M., Smits, R., Van Waveren, B.: Roles of systemic intermediaries in transition processes. Int. J. Innov. Manag. **7**(3), 1–33 (2003)
10. Dalziel, M.: Why do innovation intermediaries exist? In: Paper Presented at the 2010 Druid Conference, 16–18 August, Imperial College Business School, London, 2010 (2010)
11. Rosvall, M., Bergstrom, C.T.: An information-theoretic framework for resolving community structure in complex networks. Proc. Natl. Acad. Sci. **104**(18), 7327–7331 (2007)
12. Rosvall, M., Bergstrom, C.T.: Maps of random walks on complex networks reveal community structure. Proc. Natl. Acad. Sci. **105**(4), 1118–1123 (2008)
13. De Domenico, M., Lancichinetti, A., Arenas A., Rosvall, M.: Identifying modular flows on multilayer networks reveals highly overlapping organization in interconnected systems. Phys. Rev. X, **5**(1), 011027 (2015)

14. Righi, R.: A methodological approach to investigate interactive dynamics in innovative socio-economic complex systems. Ital. J. Appl. Stat. **30**(1), 113–142 (2018)
15. Shapira, P., Youtie, J.: Impact of technology and innovation advisory services. In: Edler, J., Cunningham, P., Gök, A. (eds.) Handbook of Innovation Policy Impact. Edward Elgar Publishing, London (2016)
16. Etzkowitz, H., Leydesdorff, L.: The dynamics of innovation: from National Systems and "Mode 2" to a Triple Helix of university—industry—government relations. Res. Policy **29**(2), 109–123 (2000)
17. Cooke, P.N., Heidenreich, M., Braczyk, H.J. (eds.): Regional Innovation Systems: The Role of Governance in a Globalized World. Psychology Press, Colchester (2004)
18. Edler, D., Rosvall, M.: The MapEquation Software Package. http://www.mapequation.org
19. Newman, M.E.J., Girvan, M.: Finding and evaluating community structure in net-works. Phys. Rev. E **69**, 26–113 (2004)
20. Fortunato, S., Hric, D.: Community detection in networks: a user guide. Phys. Rep. **659**, 1–44 (2016). https://doi.org/10.1016/j.physrep.2016.09.002
21. Kivelä, M., Arenas, A., Barthelemy, M., Gleeson, J.P., Moreno, Y., Porter, M.A.: Multilayer networks. J. Complex Netw. **2**, 203–271 (2014). https://doi.org/10.1093/comnet/cnu016
22. Arenas, A., De Domenico, M.: Nonlinear dynamics on interconnected networks. Physica D Nonlinear Phenom. **323–324**, 1–4 (2016). https://doi.org/10.1016/j.physd.2016.03.016
23. Solé-Ribalta, A., De Domenico, M., Gómez, S., Arenas, A.: Random walk centrality in interconnected multilayer networks. Physica D Nonlinear Phenom. **324–324**, 73–79 (2016). https://doi.org/10.1016/j.physd.2016.01.002
24. Sewell, W.H. Jr.: A theory of structure: duality, agency and transformation. Am. J. Sociol. **98**, 1–29 (1992)
25. Vedres, B., Stark, D.: Opening Closure: Intercohesion and Entrepreneurial Dynamics in Business Groups (October 7, 2008) (2008)
26. Brin, S., Page, L.: The anatomy of a large-scale hypertextual web search engine. Comput. Netw. ISDN Syst. **30**(1–7), 107–117 (1998)
27. Caloffi, A., Rossi, F., Russo, M.: The emergence of intermediary organizations: a network-based approach to the design of innovation policies. In: Geyer, R., Cairney, P. (eds.) Handbook on Complexity and Public Policy. Handbooks of Research on Public Policy series, pp. 314–331. Edward Elgar Publishing, Cheltenham (2015)

Inter-Organizational Networks and Third Sector: Emerging Features from Two Case Studies in Southern Italy

Andrea Salvini, Antonietta Riccardo, Francesco Vasca, and Irene Psaroudakis

Abstract Social Network Analysis is a useful technique for studying emergent behaviours of cooperation, intervention and governance in inter-organizational networks. In this work, an empirical study of two networks of organizations operating in local territories in Southern Italy and focusing on Third Sector and welfare activities is presented. The actors are committed to experimenting a model of coordinated intervention induced by two corresponding egos which are local Caritas centres. The nodes of the two graphs are determined by combining ego-network and whole-network approaches. The weighted edges representing mutual knowledge and collaboration between nodes are determined through interviews with all actors of the local groups. It is shown that metric properties of the networks can be useful indicators to monitor and evaluate endogenous features, e.g. relational and structural embeddedness, and exogenous features characterized by homophilic mechanisms. The analysis provides insights on the networks governance of the social interacting organizations and reliable descriptors of the social processes that govern their functioning.

Keywords Network governance · Social network analysis · Third sector · Inter-organizational networks · Network effectiveness

1 Introduction

The graph-based modelling of inter-organizational networks has been demonstrated to be a useful tool for the analysis of social interactions.

A. Salvini (✉) · A. Riccardo · I. Psaroudakis
University of Pisa, Pisa, Italy
e-mail: andrea.salvini@unipi.it; antonietta.riccardo@phd.unipi.it; irene.psaroudakis@sp.unipi.it

F. Vasca
University of Sannio, Benevento, Italy
e-mail: vasca@unisannio.it

© Springer Nature Switzerland AG 2020
G. Ragozini, M. P. Vitale (eds.), *Challenges in Social Network Research*,
Lecture Notes in Social Networks, https://doi.org/10.1007/978-3-030-31463-7_13

209

The aim of this article is to show how the metric properties of the collaboration networks can be useful indicators to monitor and evaluate the network governance of social interacting organizations, and to act as coherent and reliable descriptors of the social processes that govern their functioning. In this regard, we will present some methods of analysing the network structures of public and Third Sector nodes starting from two case studies. The empirical analysis is carried out by considering the networks promoted by the Caritas centres operating in the local territories corresponding to two dioceses in Southern Italy, i.e. Aversa and Benevento. These two centres are the egos of the networks, while the other nodes of the graph are determined based on some specific ongoing activities. The combined use of ego-network and whole-network strategies is proposed as a methodological approach for the identification of the network boundaries: a leading ego node is chosen by considering its high representativeness with respect to the specific network goal of interest; then all nodes of the ego-network are interviewed by using a whole-network perspective, thus determining the weighted collaboration among the actors. The weighted graph resulting from this procedure is called the "ego-whole" network.

The study of structural parameters of the two ego-whole networks will provide information on the "health status" of the analysed networks and their effectiveness. The analysis of the structural components that we propose can be considered as an essential preliminary step for studying the elements which contribute to the effectiveness of governance in this type of networks [1], and for identifying suitable actions aimed at improving the network's interactions and strategies [2–4].

The rest of the work is organized as follows: In Sect. 2 a state of the art overview on the existing results showing the usefulness of SNA for real inter-organizational networks operating in the Third Sector is presented. Then Sect. 3, by presenting some empirical findings, synthesizes why and how the recent Italian welfare system legislative reorganization indicates a central role for networked actions among social entities. Section 4 presents the two local initiatives considered as guidelines for the proposed analysis. Section 5 proposes a brief description on the existing approaches for evaluating the effectiveness of community welfare networks. The technique is then applied to the two case studies: Section 6 shows how the networks data are obtained. Section 7 describes the main results achieved; Sect. 7.1 is about the structure of the networks, related to their relational and structural embeddedness, while Sect. 7.2 proposes a similarity analysis based on structural equivalence and blockmodelling of the two real networks. Section 8 concludes the paper with the discussion of the results and the indication of possible directions for future research.

2 The Network Perspective in the Analysis of Inter-Organizational Systems

Inter-organizational networks can be defined as intentionally created relationships among at least three autonomous but interdependent organizations that cooperate in order to achieve a common result (*output*) or to jointly produce an expected emergent behaviour (*outcome*). For instance, one can think at organizations that share information and resources for counteracting certain social problems, or for promoting innovations in social, educational and environmental interventions ([5, 6]; for further information on the topic of network governance see [7, 8]).

In the Italian social context, the transformations of local welfare and the introduction of the "Third Sector Code" have recently given new impetus to the debate if and how new forms of collaborations among civil society actors would be able to face the challenges of these changes and contribute to the development of territorial communities. Within the vast universe of the Third Sector, forms of synergy and partnership have been recently tested and considered good practices to be promoted since they generate significant results in terms of sharing resources and coping with local needs. Despite the ongoing fragmentation and diversification of regional and local welfare systems, the idea that the expansion of the "public sphere" can positively influence the effectiveness of public policies in the social, health, cultural and environmental fields has been consolidated [9–11]; this public sphere obviously requires a virtuous collaboration between the various actors operating on the territory under the forms of partnerships and networks of different kind of organizations and institutions. This awareness does not derive only from the well-known reduction in resources allocated to the financing of interventions in the public dimension, but also from the need to direct the action of welfare systems on the basis of value and functioning frames that are consistent with the complexity of the phenomena (such as in the case, just to give an example, of the so-called generative welfare or even of community welfare) [11–13]. In these new contexts, one of the most important issues undoubtedly consists of understanding the forms of institutional integration that allow those objectives and, above all, whether the various actors are aware of and prepared to face the challenges of this networked substantial reform of the local welfare.

The Title VII of the new "Code of the Third Sector", art. 55 and 56, introduces some fundamental principles in the regulation of relationships between Third Sector entities and public institutions; in particular, it is expected that the latter will ensure the active involvement of the Third Sector entities, through forms of network governance. The new rules will clearly require significant changes in the implementation of the local welfare systems. On the other hand, the new regulatory framework represents an opportunity to systematize those co-planning experiences that have already been developed for some years, albeit in a fragmented way, in many areas of the Country. Such experiences should no longer constitute meaningful episodes by their "exemplarity", and they should become widespread practices in the relations between Third Sector organizations and these institutional entities [14]. The use of the concepts of *network*, *network intervention* and *network governance* has been a rather widespread refrain in many local welfare systems; however, in most areas this reference remains at an essentially evocative level without leading to widespread and shared network strategies, based on theoretical and methodological awareness. The rare studies carried out on this issue in Italy show a considerable difficulty, if not a real reluctance, by the Third Sector organizations, to promote and develop forms of partnership that can effectively be considered as network-governed partnerships [15]. However, it is possible to find experiences around the Country that describe a significant effort by the most sensitive Third Sector organizations. They promote a vision for which the construction of collaborative systems which involve institutional entities and the Third Sector can generate virtuous effects both within the same networks and towards the served community. The growing literature on social capital (among others [16]) and, more specifically, on network governance [17, 18] is an indicator of this sensitivity.

Following Raab and Kenis [6], it is important to distinguish between two types of network: the first type refers to a structure of relationships of informal interdependence between collective social actors that "emerges" from their dyadic interactions, without further specification of objectives, timing and ways of "being together"—a mode that does not necessarily create awareness and collective identity in the members. The second type, on the other hand, is more concerned with network governance, whereby the systems of interdependence between collective entities are intentionally created to achieve certain goals—modalities that instead envisage the construction of specific identity frames.

A central question in this debate is how the effectiveness of network governance can be assessed [19–21]. The effectiveness of a network can be defined as "the attainment of positive network-level outcomes that could not be achieved by individual organizational participants acting independently" [7, p. 4]. Provan and Milward [1] correctly observe that the assessment of network effectiveness must necessarily take into consideration a plurality of aspects, such as the impact of network activities in the served community, the role of individual members and the nature of the interactions between them, the evaluation of the stakeholders and, obviously, the characteristics of the structure of the interaction network. Social Network Analysis (SNA) is an appropriate and comprehensive methodological perspective that can help in studying the effects of the network structure on governance and outcomes, as well as in supporting the management of these governance processes [8, 22]. SNA makes available to scholars and practitioners a set of methodological tools and techniques to verify if and how the structural configuration of those relationships generates effects both on the governance of the network and on its outcomes [23, 24], thus encouraging the involvement and empowerment of collective actors operating at the level of local communities [25, 26]: the intuition on the importance of working together and sharing resources can be anchored through SNA into a solid and validated conceptual and theoretical framework.

3 Networks and Welfare Systems in Italy: Some Empirical Findings

The literature on empirical experiences of inter-organizational networks realized in Italy in the framework of local welfare systems is anything but extensive. Most of the analyses consider case studies limited to specific territorial contexts. Among these experiences we can mention a study on the propensity of some Italian Third Sector organizations—especially social enterprises—to build innovation networks with the aim of evaluating their performance compared to the level of cooperation in the network [27]. The authors, by adopting SNA techniques, argue that the highlighted innovation is the result of the nodes interactions inside the network, which, unlike what one could expect, has a rather low level of cohesion (measured through the usual density parameter). Furthermore, it is pointed out that innovation networks are more effective when the nodes adopt an "open" approach for building new connections with entities that are outside the formal boundaries of the partnership. SNA techniques have been applied in the recent work of Delle Cave for studying formal and informal networks between Third Sector organizations in the Municipality of Naples [28]. The analysis therein verifies that the outcomes of local welfare policies are not much connected to the "performance" of the single actors; indeed, the key element is represented by the development of collaborative practices among a multiplicity of interdependent actors that exchange different kinds of resources in order to face needs of collective relevance. The adoption of the SNA framework allows one to identify significant relational areas within the network such as those of the centre and the periphery [29], to evaluate the most relevant and active actors (the "core" of the network), the "structural holes" present in the structure [30], the cliques and the actors who assume a leadership role in the development of the sector. The practice of network collaboration represents a strategic step in the development of the sector, especially in scenarios where a gradual and massive reduction of resources is present; however, the results of the research reveal the importance played by some specific "nodes", compared to others, in the processes of construction, maintenance and development of networks, according to their economic and organizational strength. Delle Cave points out that consortiums of cooperatives represent the main protagonists of the evolution of the Neapolitan Third Sector, also because of their propensity to cultivate a specific competence in local and supra-local networking. Similar conclusions were already proposed by Corbisiero [31] who highlighted the multi-centred structure of the network describing the informal relationships among Neapolitan Third Sector organizations. Moreover, the author also demonstrates that the network characterized by formal agreements among the nodes clearly shows the high centrality of few nodes which are the most structured organizations.

By adopting the SNA perspective, Salvini [32] analyses the relationships of knowledge, exchange and collaboration between voluntary organizations operating in three different territorial areas of Tuscany. The "structural" network parameters reveal a rather singular situation, which denotes a clear self-referentiality of the Third Sector, with a very low level of density, particularly in the exchange and collaboration networks, a high level of centralization, and a low level of reciprocity and transitivity. Moreover, the high levels of centralization reported in the survey outline a clear dependence of many organizations on the resources exchanged with a limited number of nodes, which instead enjoy a position of privilege and centrality. These studies show the usefulness of SNA for interpreting the structure and the relational dynamics that characterize the experiences of partnership and collaboration among Third Sector organizations and other public and private players. The existing literature shows that such networks are essentially characterized by low levels of density and reciprocity, and a high level of centralization—which describes the polarization of network relations around some "strategic" nodes that play the role of "key players" in the "control" of the flow of resources. Although the characteristics outlined here are not found in all networks constituted by these classes of actors—and with the same intensity—it is evident that these features reduce the networks effectiveness, especially from the point of view of their governance.

The analysis of the networks developed in the collaboration partnerships described above allows one to gather useful elements for the analysis proposed in this work which focuses on the networks of collaboration between two sets of local actors in two contiguous but different territorial areas.

4 The Collaboration Networks of the Caritas in Aversa and Benevento

The two egos of the corresponding networks considered in this article are the Caritas of the Diocese of Aversa (RCA) and the Caritas of the Diocese of Benevento (RCB), both active in Southern Italy. Two elements justify the choice of type and location of the selected case studies. Firstly, the national Caritas organization has recently stimulated a general reflection on the importance of improving local networked organizational forms in order to respond in a synergic and coordinated way to the social needs of the served communities, thus overcoming the fragmentation and self-referentiality of the local welfare agencies. Furthermore, the choice of the two inter-organizational ego-whole networks is motivated by recent efforts spent on these territories for experimenting a model of intervention founded on the valorization of Third Sector community networks [33, 34]. More specifically, the two local groups of Caritas have set themselves the goal of generating inter-organizational networks, promoting exchange projects and collaborations involving Third Sector entities, local governments, educational institutions and church organizations. In this generative process, the Caritas of the two dioceses have played a role of animation and coordination, aimed at promoting the creation of networks according to the different project opportunities that were created on the territory. The characteristics of the two networks, however, differ because of the dissimilar cultural and organizational contexts in which they operate, the peculiarity of the issues identified as targets for their actions, and the governance style adopted.

The two networks have been promoted and supported by egos belonging to the ecclesial community whose nature and organizational structure are characterized by specific references in terms of vision and mission. Then, it is interesting to point out that within cultural frames in which hierarchical relationships play a significant role, clear awareness has matured about the strategic nature of building networks, as well as open and widespread cooperation in order to efficiently achieve specific objectives and appropriately serve the territorial community. Such awareness is even more significant if we consider that the international literature on SNA exhibits very rare evidence regarding situations of network experiences and interventions that directly involve churches and church organizations. One of these studies considers SNA for evaluating the experimental participation of two US Catholic parishes in the implementation of health programs and interventions in two communities of Massachusetts [35]. The results showed that the networks which involved more volunteers and were characterized by a lower level of centralization (for example, less polarized on the role of the parish priest) realized a more effective achievement of the intervention's objectives.

Another study focused on the social capital generated within some ecclesial congregations in Australia, which was compared with the social capital generated in the broader social community where those congregations operated [36]. These studies have shown how faith communities can be significant repositories of social and relational capital and that the exploitation of such wealth is strictly related to the configuration and properties of the relational structures (networks) through which they operate [37].

5   Assessing Inter-Organizational Networks

Different approaches can be used to evaluate the effectiveness of community welfare networks. Three main mechanisms which relate the analysis of structural parameters to the network action can be identified: two of them are endogenous (*relational embeddedness* and *structural embeddedness*), while the third is exogenous (*relational homophily*). The two endogenous mechanisms were identified by Granovetter [38] and reported by other authors afterwards [20, 39]. By considering the purposes of maintaining and developing the network, the former endogenous approach reveals the importance of the strength of the dyadic bonds (measured through *reciprocity*), while the latter concentrates on the relevance of the dynamics of composition and interdependence between the various subsystems of the network (measured through *transitivity* and *clustering*). From the results of these researches, it emerges that the organizational design of governance networks—and therefore the governance "style" adopted by the members, as well as the level of informality/formality of the network—can positively influence these mechanisms and the corresponding indicators, generating desirable structural features. For example, the absence or lack of reciprocity effects in the network are indicators of a circumscribed level of "relational" embeddedness, i.e. a sign of inadequate levels of dyadic collaboration. However, the lack of transitivity effects and triadic exchanges can reduce the capacity for mutual influence, to establish multiple social bonds within the network, and ultimately to build what the authors define "structural" embeddedness, which is a "distributed" collaborative context and goes beyond dyadic relationships [20]. The exogenous mechanism of homophily, instead facilitates and promotes the creation of stable dyadic relations and the "closure" of areas inside the network, although based on criteria of similarity related to the attributes of the nodes. The effects of homophily may be desirable in order to generate areas of greater connection in the network, but at the same time they can also generate homogeneous clusters that are not able to convey differentiated network resources, if not properly connected to each other.

These mechanisms will be considered in our analysis for assessing whether and under which conditions they are present in the two cases of study. More in details, the main objectives of the study concern the analysis of the structure of the two inter-organizational networks with the network governance perspective and the analysis of the role played by the two Caritas organizations as promoters for the development of the corresponding networks.

The three structural mechanisms defined above will be detected by using the level of relational embeddedness and structural embeddedness as indicators of endogenous effects in network operation processes. More specifically, high values of the two kinds of embeddedness are interpreted as indicators of a network configuration that promotes collaboration between nodes and effectiveness in governance (regardless of the specific outcomes of the interventions). The presence of relational homophily will be also detected, a mechanism whose activation exerts a significant influence on the network functionality, which does not always or necessarily correspond to positive connotations.

Regarding the role played by the egos in the networks analysed, it is worth mentioning that previous works dealing with inter-organizational networks in Italian social contexts showed the typical presence of nodes with high centrality sometimes responsible of a differentiated access to resources for the other nodes. Krebs [3] and Anklam [2] argue, on that point, that the role of these "hubs" for a correct network development is not to polarize the system of relationships around oneself and to one's own action, binding or influencing in this way the action of the other nodes to one's choices, but to favour the realization of multi-hub configurations, in which the structure of relationships is widespread and characterized by complementarity. In the present work we will show if and how the network actions of the two Caritas nodes are aimed at intensifying the links and "delivering" to other nodes the role of promoters, thus encouraging the achievement of ever-increasing levels of connection and exchange of resources.

As a side effect of the study proposed in this article, by connecting the analysis of the two ego-whole networks with the three mechanisms we will provide a SNA perspective which could be potentially useful for the analysis of similar community welfare networks in Italy.

6   The Research Plan: Ego-Whole Network Construction, Data Collection and Analysis

The analysis of the two Caritas networks is carried out through the development of three consecutive phases:

- Identification of the nodes of the two networks under investigation;
- Data collection;
- Analysis and interpretation of data.

The first essential step consists of determining clearly the boundaries of the network by identifying the nodes that are part of it; we proceeded through a combined strategy of *ego-network* and *whole-network*. The Caritas of the two territories were considered as the ego node of the network, which was asked, through the name generator technique, to indicate the other nodes ("alters") that are part of their network, current or potential, that is, those organizations or entities to eventually involve into project activities. The boundaries of the network, therefore, are established by the ego (the Caritas), which we have called "node generator". In this way we were able to build a first-level network, composed of ego and all the elicited alters at a distance of one (weighted) link from ego. Following the most consolidated data collection practices in ego-network mode processes, at this point the researcher usually asks ego, through a technique called "network interconnector", to indicate if—based on their knowledge—there are links between the elicited alters—so as to reconstruct the configuration of the relationship structure between the alters. This way of constructing the (ego)network—which is the most widely used due to its economy—is also one of the most evident limits of the ego-network technique, given that the networks obtained are essentially the outcome of an ego cognitive process, which obviously introduces bias and distortions that are not completely controllable [40]. To overcome this limit, the construction phase of the network was completed by submitting the knowledge questionnaire to all alters organizations elicited by the ego. They were asked to specify the existence and intensity of the link with each of the alters, choosing one of five increasing levels of "strength" of the relationship: no knowledge, lack of knowledge, good knowledge, relationship of exchange, cooperation. The resulting directed graph is what we call "ego-whole" network.

"Ego" contributed to define the boundaries of the network, but the relational system has "emerged" through the interviews carried out with each individual alter, according to what is commonly done in whole-network investigations; consequently, it was not necessary to use the *name interpreter* (the tool for the collection of attribute data, that is, of the characteristics, of the individual nodes), as it is normally performed in surveys with ego-centred techniques. The only bias that must be taken into account is, of course, the presence of at least one node with a degree equal to the size of the network, i.e. the Caritas nodes. However, it has been preferred not to exclude the "generating nodes" from structural analyses since the distortion performed from above outdegree of the Caritas nodes is offset by the

Table 1 Composition of RCA and RCB networks by type of organization (category), distinguished by their ego whole-networks and the number of nodes which answered to questionnaires (answers)

Abbr.	Category	Caritas network Aversa (RCA)		Caritas network Benevento (RCB)	
		Total	Answers	Total	Answers
AS	Associations	19	11	12	10
CP	Cooperatives	1	1	7	7
OS	Second-level organizations	0	0	1	1
EM	Religious organizations	1	1	2	2
GI	Informal groups	4	4	0	0
ES	Social-health care institutions	6	6	3	3
SU	Schools–universities	7	4	6	6
AL	Others (parishes, private organizations)	6	3	9	6
	Total	44	30	40	35

procedure adopted for the network construction, which obviously take for granted the existence of reciprocation in the considered relationship.

The representatives of the Caritas of Aversa and Benevento were asked to indicate which institutional entities, Third Sector and ecclesial organizations they would eventually contact for carrying out intervention and project activities in the context of the territorial community welfare. The ego-whole network generated by the ego Caritas of Aversa (RCA) consists of 44 nodes, while the network rising from the ego Caritas of Benevento (RCB) consists of 40 nodes in the case of RCB. The data gathering phase was developed through questionnaires to egos and alters during the semester between April 2017 and September 2017. Table 1 shows the composition of the networks by type of organization. The two networks are characterized by a high presence of Third Sector associations and educational and socio-health institutions, with an interesting difference between the two territorial areas relative to the presence of social cooperatives, which are more consistent in the RCB. The answers received to the questionnaires correspond to 75.0% of the RCA nodes and 87.5% of the RCB nodes.

It is useful to highlight some differences between the two networks. In RCA the organizations operate mainly within restricted territorial areas, often coinciding with the district or with the country in which the legal or operational head offices are located. The network is consolidated mainly around local initiatives, and its mobilization takes place through sporadic participation in Caritas activities and specific interventions of solidarity and voluntary work. The organizations of RCB have widened the range of their interventions throughout the diocesan territory, demonstrating greater involvement in the proposed reticular initiatives. These nodes are active in searching for relationships with differentiated territorial partners and in sharing resources and skills for accessing funding channels provided by national and international institutions. Furthermore, in RCB there are more social cooperatives than in RCA, and a second-level organization is established, consisting of a consortium of social cooperatives; in general, RCA is characterized by less structured organizational coordination. In both networks, organizations active for more than 10 years and with a medium-high level of committed human resources prevail.

7 Main Results

7.1 The Structure of Networks: Relational Embeddedness and Structural Embeddedness

Figure 1 shows the directed graphs corresponding to RCA and RCB. It is interesting to note that the interviews showed the awareness, by most of the alters, of belonging to the corresponding Caritas network. This confirms the proper definition of the networks boundaries which included the most significant informal relationships that connect the pairs of nodes.

In Table 2 the descriptive parameters of the two networks are illustrated and classified by the type of relationship between the various nodes. Since the graph is directed, the number of potential ties has been considered to be $n \ (n - 1)$, where n is the number of nodes of each network. The structural measures show some substantial differences between the two networks in terms of the presence and strength of the ties.

Considering all types of relationships, including those of lower intensity, the density is equal to 0.58 for the Aversa network and 0.75 for the Benevento network; especially for RCB, the density assumes rather high values. The centralization index is higher where the density is lower, particularly in RCA, where relations therefore depend more on the activity of a limited number of nodes. The values of these global properties are sensitive to the different types of ties between the nodes, so as evident by considering the merged ties classes reported in Table 2. Density and centralization also depend on the different binding classes. The density significantly decreases by considering the ties based on exchange, collaboration and good knowledge, while the centralization grows in correspondence to the decrease of the density. From the analysis of these data it is possible to conclude that the two networks have characteristics, albeit in different intensity, which confirm a typical feature of relationships in Third Sector networks. We are in the presence of high levels of mutual knowledge, but also of low levels of collaboration and exchange, which are usually polarized around the activities of a few organizations—generally the most structured and consolidated from the point of view of the management of human and economic resources. However, there are further observations to be done with specific reference to the analysed networks. The degree of reciprocity in the network, if we consider all the links, is quite high: 0.63 for RCA and 0.80 for RCB; this is a sign that one, albeit generic, mutual knowledge of the organizations on the territory is consolidated. However, these values tend to decrease when evaluating

Table 2 Descriptive parameters of RCA and RCB networks, classified by (a) the type of relationship between the various nodes (lack of knowledge, good knowledge, exchanges, collaboration), (b) the combined relationship between the same nodes (collaboration + exchanges, collaboration + exchanges + good knowledge), (c) the whole-networks (lack of knowledge + good knowledge + exchanges + good knowledge)

	Lack of knowledge	Good knowledge	Exchanges	Collaboration	Collaboration + exchanges	Collaboration + exchanges + good knowledge	Whole-network
Caritas network Aversa—RCA							
No. of ties	190	80	49	154	203	283	473
Density	0.23	0.09	0.06	0.19	0.25	0.34	0.58
Centralization	0.28	0.19	0.09	0.69	0.74	0.67	0.43
Clustering	0.28	0.17	0.03	0.50	0.52	0.52	0.65
Reciprocity	0.22	0.10	0.16	0.51	0.56	0.55	0.63
Caritas network Benevento—RCB							
No. of ties	320	150	115	313	428	578	898
Density	0.26	0.12	0.09	0.26	0.36	0.48	0.75
Centralization	0.27	0.17	0.08	0.49	0.54	0.53	0.25
Clustering	0.32	0.20	0.11	0.59	0.61	0.62	0.79
Reciprocity	0.32	0.19	0.09	0.58	0.62	0.64	0.80

Fig. 1 RCA (up) and RCB (down) networks

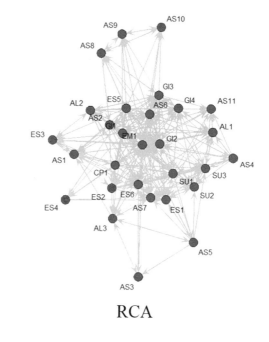

RCA

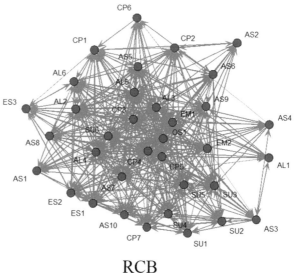

RCB

only the relationship based on good knowledge and the exchange of information and resources while decreasing less if we consider the ties based on collaboration. The data related to reciprocity is particularly important to understand the strength of the ties at the dyadic level (and hence the measure of *relational embeddedness*) and the functioning of the system [20]; a low degree of reciprocity indicates a differentiated access to the resources mobilized and a low propensity to the mutual use of those resources. The combined analysis of the parameters shows that the two networks (especially for RCA) have "unequal" access to resources, above all due to the dependence of many organizations on the resources made available by a few nodes, which in turn do not need to draw on the resources of others. In addition, a low degree of reciprocity is an indicator of poor sharing and agreement about the contents and objectives that drive the relational structure, as well as the rules that should inform the whole-network [20]. The level of agreement and awareness of those contents, objectives and rules is an important indicator of the functioning of the governance system, although the analysed networks are based more on informal relationships than on formal agreements. In both situations we can observe a *weak degree of relational embeddedness*, not so much on the level of an abstract awareness of the importance of collaborating to achieve collective goals, but regarding concrete experiences of cooperation based on mutual exchange of resources. It can therefore be considered that the reference to "collaboration" is more of an evocation, also ethically oriented, than a concrete experience of mutual sharing of resources and competences aimed at achieving common goals. From this point of view, the reference to the exchange appears obviously more coherent with the mathematical parameter of reciprocity. It is not by chance that the level of reciprocity in the exchange—however strange it may seem—is very low. This does not mean that there are no resource flows in the network, but that these flows do not provide for reciprocation mechanisms. The degree of *structural embeddedness* can be analysed through the measurement of the *clustering coefficient*. The level of reciprocity it could lead one to expect a not particularly positive performance with respect to clustering. It is possible, on the other hand, to observe that both networks have a very high global coefficient, which remains robust in the relational structures marked by "good knowledge" and "collaboration"— while it is reduced, as expected, in "exchange" relations. The high level of the clustering coefficient suggests that the structure of the two networks is sensitive to the formation of "clouds" of nodes based on some aggregation criterion. It therefore emerges that for the two networks, the degree of structural embeddedness has a consistency, which however is decreasing in the relationship subsystems more concretely founded on the exchange of resources and information. In other words, this means that the potential for mobilization—and cohesion—is high, unlike what has been experienced in the two networks in terms of effective experiences of sharing resources in non-dyadic forms. At the global structure level, the presence of a high clustering could favour—following the most well-known theories of social capital and structural holes—mechanisms of "closure" of the many open triads, and therefore the achievement of higher levels of cohesion of the network. This eventuality certainly appears to be desirable if the network governance concerns projects aimed at achieving specific goals and relations systems legitimated by formal agreements. In situations such as those analysed, in which networks are mainly informal networks to be activated gradually based on the specific project opportunities, it is more desirable to maintain a "hub and spoke" or "centre-periphery" type structure, as reported by several authors. A high level of clustering (as in the case of the two networks analysed) is certainly useful for achieving these forms of governance, provided that the centralization rate is not too high. In general, RCB shows characteristics that are more consistent with a "hub and spoke" functioning of governance, while in RCA the structure is even more dependent on the activity of a few organizations, including the promoter (Caritas Aversa). These are informal networks certainly founded on the willingness to collaborate, but in which the practice of that collaboration in a case depends essentially on the organizational role of some specific entities, in the other on a coordinated and reciprocal action of a higher number of organizations.

An evident characteristic of the two networks, especially for RCA, is the high degree of centralization of the network, which is a symptom of the tendency towards the polarization of collaboration relationships around a limited number of nodes. This means, in turn, that the distribution of the degree of the nodes shows considerable variations, which deserves a more extensive analysis. By focusing our attention to the five nodes that present the highest centrality indegree—excluding the two "generating nodes"—we can see that in RCA they are public institutions (schools or social-health services) and national associations with local agencies, while in RCB they are social cooperatives.

On the other hand, if we consider the five nodes with the highest outdegrees, we find organizations of a similar nature, mainly local associations and church offices in RCA and social cooperatives in RCB. It is also useful to observe which are the nodes that present a consistent difference between indegree and outdegree centrality: the most popular but less active nodes (high indegree and low outdegree) are national associations (therefore very well known in the local area and connected to the Caritas networks only for specific activities) and organizations belonging to the School/University category. The youngest organizations in the networks are among the most active but less popular nodes (high outdegree and low indegree). Another node of this type corresponds to an organization whose manager is also referent of other organizations—and therefore constitutes a *trait d'union* among many actors. In RCB, social cooperatives are particularly significant both in terms of indegree and outdegree, and are dynamic entities characterized by considerable practical initiatives. By classifying the node centrality within the sub-networks related to different intensity (and type) of relationships, the results shown in Table 3 highlight that the first three nodes are also listed as those with the highest degree in each of the four relationship intensity classes.

Beyond the confirmations related to the presence of specific organizations with high degree within the sub-network of relationships characterized by lack of knowledge, it should be noted that in the Aversa network the most dynamic nodes (for activity and popularity) in terms of collaboration are entities belonging to the church field, while in the Benevento network they were born by "budding" from the Caritas node. This element suggests that cooperation develops more easily between similar nodes, because of homophily mechanisms. The construction of the network structure is sensitive both to the presence or absence of endogenous factors, such as reciprocity and clustering, and to exogenous ones, such as the level of similarity of the node attributes. These aspects indicate a level of network

Table 3 Classification of the nodes having the higher degree centrality related to different types of relationships (sub-networks)

Sub-networks	Caritas network Aversa—RCA		Caritas network Benevento—RCB	
	No. of ties	Nodes with higher degree centrality	No. of ties	Nodes with higher degree centrality
Lack of knowledge	190	RCAAS10 (24)	320	RCBAS2 (33)
		RCAAS1 (21)		RCBCP2 (28)
		RCAES1 e RCAAS11 (19)		RCBAS3 (28)
Good knowledge	80	RCAAS6 (16)	150	RCBSU6 (21)
		RCACP1 (15)		RCBAS7 (17)
		RCASU1 (12)		RCBAL2 (16)
Exchanges	49	Caritas (13)	115	RCBOS1 (17)
		RCAAS7 (8)		RCBCP4 (15)
		RCAES1 (7)		RCBAL4 (13)
Collaboration	154	Caritas (42)	313	Caritas (52)
		RCAGI2 (23)		RCBCP5 (45)
		RCAGI1 (22)		RCBCP3 (36)

development that is not yet particularly "mature" towards the achievement of the aforementioned "hub and spoke" composition [3]. Therefore, in order to better understand the operative conditions of the governance networks it is useful to provide a deeper analysis of the similarities of the nodes. To this aim, the collected data have been elaborated by using the blockmodelling technique and the structural equivalence. This type of analysis will provide information about the presence of clusters between organizations possibly characterized by high levels of similarity in terms of their positions in the network.

7.2 Structural Equivalence and Blockmodelling for the Analysis of Similarity

Homophily is usually referred to similarity based on specific individual characters (attributes) of the nodes. The identification of such characters for the inter-organizational homophily is difficult to accomplish, since the nodes could be similar for a specific aspect, but very different from others. Therefore, it is useful to use the blockmodelling technique in order to analyse the way in which the nodes of the network are distributed within subsets according to the mechanism of structural equivalence [41]. Two nodes are defined structurally equivalent if they exhibit the same configuration related to incoming (indegree) and outgoing (outdegree) ties. A "block" consists of a set of network links that connect pairs of the subsets thus constructed; in this way it is possible to anchor the similarity to the structural position covered by the nodes within the network. From an interpretative point of view, the nodes belonging to the same sub-group or group are subject to similar opportunities or constraints of a structural nature—and can be considered structurally homophiles.

The blockmodelling analysis has been carried out by deriving the binary adjacency matrices of the two networks. In particular, the relations "lack of knowledge" and "good knowledge" have been merged by assigning to these relations the value 0, i.e. no link, while the ties "exchange" and "collaborate" have been associated to unitary links. The binary matrices allow us to concentrate on the existence of links with highest intensities between the nodes, i.e. those described by exchange and collaboration relationships. Structured equivalence was measured by using the CONCOR (CONvergence of iterate CORrelation) algorithm as a clustering method. Figure 2 shows the block matrices of the Aversa and Benevento networks with the corresponding clusters of structurally equivalent nodes. In the case of RCA, four blocks of ties are identified, while three blocks of ties in the case of RCB, which indicates a lower level of differentiation in the Benevento network. Let us shortly analyse the composition of these blocks. As it was expected, in RCA the Caritas node constitutes a block in its own, an aspect that confirms the relevant role of the "generating node" within the network. Blocks #2 and #3 are the densest areas of interactions, particularly between #2 and #3 and between #3 and #4; block #4

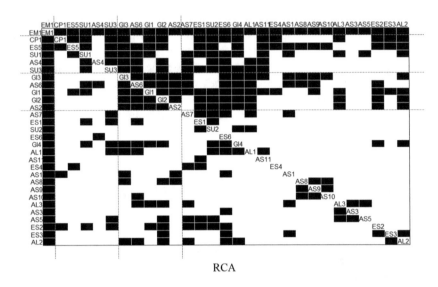

RCA

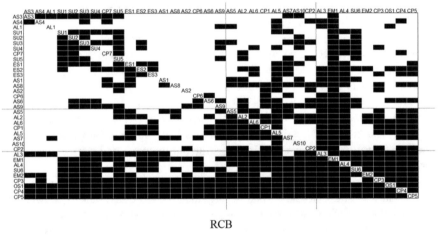

RCB

Fig. 2 Blockmodels of RCA (up) and RCB (down) networks, created by using CONCOR (CONvergence of iterate CORrelation)

is the largest in terms of nodes but also the least dense in terms of interactions between nodes and between blocks. By observing the single blocks, it is possible to detect different internal levels with respect to the type of nodes. More specifically, block #2 is formed by quite heterogeneous entities (a cooperative, a social-health care institution, two schools/universities, an association), while block #3 consists of three informal groups and two associations, therefore relatively similar entities; block #4 collects all the other nodes, consisting essentially of associations, social and health care institutions, educational institutions. The "core network" of RCA, therefore, consists of the generator node, and two sets of nodes that are certainly heterogeneous with each other and with a different degree of internal homogeneity: weaker in one and more marked in the other.

By considering the RCB network, the three blocks are internally more numerous than those of RCA. More importantly, the "generator node" of RCB does not constitute a block, but belongs to block #1, which corresponds to the block with highest density of interactions both between the nodes belonging to it and the other two blocks. In this subsystem there are four organizations that show a very high level of internal and external connections similarly to that exhibited by the Caritas node. Consequently, the role of promoter of the network is shared between different organizations, although the latter have characteristics that make them very close to Caritas itself, being three social cooperatives and a religious moral entity. A second-level organization, two parishes and a school belong to the same block. The second block consists of 5 entities of the Third Sector (two cooperatives and three associations) and 3 private entities (including the parishes); block #3, the one that exhibits the lowest density of internal and external relations, includes public institutions (educational and social-health care) and Third Sector organizations.

A compact representation of the relations between blocks is evident from the "reduced" blockmodels illustrated in Fig. 3 for both networks. The nodes of these "reduced" networks are the blocks and the links indicate the existence of relations between the blocks, i.e. between some nodes of the corresponding blocks. The thickness of the lines is proportional to the density of the ties: a thicker line indicates the presence in the original network of a larger number of ties connecting the organizations of the two blocks. The lines do not have a direction, thus showing the existence of some level of reciprocity. In the case of RCA, most of the relational activities focus around block #1 and branch off to block #2, #4 and #3; the residual activities are placed in the paths between blocks #2 and #3, #2 and #4 and #3 and #4. In the case of RCB, most of the relational activities are concentrated between blocks #1 and #3 and between blocks #1 and #2. The structure of the two "reduced" blockmodelling networks indicates, first of all, that the relational triads that connect

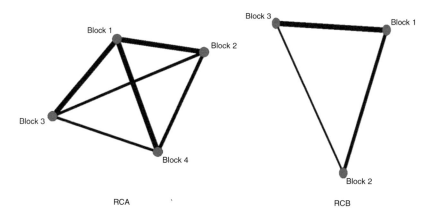

Fig. 3 Reduced blockmodels of RCA (left) and RCB (right) networks

the various nodes (blocks) are closed, a circumstance that however denotes the existence of connections between all the blocks of the network. Moreover, the block containing the generating nodes is particularly strategic for the holding and development of the two networks, particularly in RCA, since the most important lines of activity branch off from block #1. In RCB, as already noted, the smallest number of blocks and the triadic structure of the relations show a greater diffusivity of interactions.

The homophilic mechanisms act in different ways in the two networks. In RCA the role of the "generator node" is strategic in "keeping together" the various subsystems of the network, which differ from each other by their degree of internal heterogeneity of the nodes. The presence of relationships of exchange and collaboration between different entities is legitimated, therefore, not because of belonging to the same class of organizations (by legal nature or by sector of activity) but based on criteria that are independent from the attributes of the nodes, such as having taken part to networks built for specific project initiatives. The role played by the generating node in terms of "network holding" in the RCA network in RCB is covered by an entire block of organizations and relationships, which constitutes the "core network" of the overall network. This circumstance allows us to make two further considerations. Firstly, the Benevento Caritas node operates such that the responsibility for the "tightness" of the network is shared with other nodes, thus promoting a network configuration which is similar to a "hub and spoke" model. Secondly, the nodes to which this responsibility has been "transmitted" are closely related—substantially similar—to the Caritas node. In the RCA network, the homophily mechanism seems to operate independently of the attribute characteristics of the individual nodes, but based on the events (previous and present) that characterize the relationship system (for example, the fact of having shared the same design experiences). Instead, in the RCB network the homophily mechanism operates in terms of greater diffusion of the relational activities between different nodes, which nevertheless constitute in some cases a derivation by budding of the generating node. In this case it is more difficult to conjecture that homophily is independent of the individual attributes of the nodes, or at least some of them.

8   Discussion and Conclusions

Two inter-organizational networks developed within a "community welfare" experience in two territorial areas of the Campania Region in Italy have been analysed through SNA techniques. It has been shown that the SNA perspective offers an adequate theoretical and methodological tool for understanding how the structure of the relationships influences the behaviour of the inter-organizational networks governance. The international literature on inter-organizational networks operating in welfare contexts is not so vast and just few practical studies have been proposed regarding Italian territories. The analysis of the empirical experiences proposed in this paper is aimed at contributing to the construction of formal procedures for the exploitation of network parameters in the evaluation of the so-called "network effectiveness" of systems of relations among organizations active in the field of community welfare. The usefulness of organized second-level forms (partnerships, synergies, networks), more or less formalized, in order to achieve the objectives of social and health-care welfare at local level is a consolidated awareness. The creation of this type of networks has been further stimulated in Italy by the most recent social legislation—including that relating to the Third Sector Code, which encourages public administrations and Third Sector entities to practice co-design and co-planning in social interventions. The adoption of forms of distributed governance within inter-organizational systems appears to be a fundamental strategy for achieving those objectives—a situation already active in many areas of the country, although not very known. The development of these heterogeneous inter-organizational networks implies the necessity to involve specific skills in terms of managing network governance and assessing network effectiveness. This type of evaluation is a complex and certainly multi-factorial operation. In this framework the study of the structure of inter-organizational relations is a strategic element to understand the network governance and for identifying structural properties that more than others have the ability to "enable" effective governance.

The study described in this paper introduces two main original contributions with respect to what is consolidated in the literature dedicated to the topic [35, 36], and especially with reference to the Italian context. The first innovation rising from the article is methodological: for the identification of network boundaries we proposed a method which combines the classical ego-network and whole-network strategies, thus leading to what we called "ego-whole networks". At the best of our knowledge, this approach is original for the type of investigations considered, at least in the national panorama. The ego-whole network allows one to overcome the limitations of ego-centred analysis through a follow-up inclusion of the alters indicated by the "generator node" and by reconstructing the boundaries of the network through the usual modes of whole-networks. The second innovation that emerges from the analysis proposed in this paper is the demonstration that parameters usually adopted for the analysis of inter-organizational networks can be included in a simple but effective conceptual framework, which allows a more accurate interpretation of the sense of empirical SNA parameters—such as density, centralization, centrality, reciprocity and transitivity. Following, in particular, Robins et al. [20] three theoretical-methodological areas have been considered: the first two refer to the endogenous factors that contribute to the realization of the network and to the achievement of better performances in the network governance, which concern with the *relational embeddedness* and the *structural embeddedness*. The third one refers to the most recurrent exogenous factor in network theorization, i.e. the presence of some homophily mechanism in the preferential connection processes between the nodes of the network. The ego-whole network technique allows one to better interpret the role of the "generator node", highlighting how the structural position of that organization can influence the definition of the network configuration and the functioning processes. The analysis of two case studies showed that the two networks exhibit, albeit in differentiated modes and intensities, a weak degree of relational embeddedness, measured through the degree of reciprocity. Moreover, it is pointed out a more consistent level of structural embeddedness, measured through the coefficient of clustering and transitivity, although by weighting the potential mobilization of the network more than the actual experiences of collaboration and exchange of resources. The structural configuration of the RCA network is undoubtedly characterized by the role of coordination played by Caritas, oriented to activate clusters of nodes to some extent already connected to each other for some reason—such as having experienced in the past actions of mutual collaboration. The RCB network structure, on the other hand, appears less centralized and more widespread, although the nature of the nodes that have supported Caritas as a hub is very close or similar to that of the "generative node".

For a detailed analysis of homophilic mechanisms, the blockmodelling technique was adopted, which allowed us to investigate the clustering aggregation within the two networks. The RCA presents a broader clustering within a governance framework where the coordination is managed by a single hub, while the RCB presents a less clustered structuring, in which there are several organizations that are structurally equivalent in the coordination functions. It emerges, however, that these organisms are very similar to the Caritas node in their nature, thus reducing the opportunities of creating a richer and more differentiated social capital. Although structurally the multi-hub characters in RCB is more evident than in RCA, there is the risk that in the former network mechanisms based on nodal similarity are reproduced. The strategies of development and improvement of the functionality of the two networks are therefore different based on the structural differences that emerged. The immediate goal of network governance should be to reduce the level of dependence of the nodes from the coordination action carried out by Caritas and to distribute this mobilization capacity to other players, which can connect and integrate the different subsets (blocks) that have emerged thanks to the analysis of blockmodelling. Such a strategy would lead to enhancing the tendency towards the clustering of RCA through the work of various hubs—possibly differentiated from one another—aimed at connecting the different clusters to each other. In RCB, it would be necessary to proceed towards a re-formulation of the roles of the already numerous hubs present in the network, by enforcing organizations with less homophily with respect to the current hubs to gradually assume roles of coordination and activation. An essential point in the growth and maturation of networks links in the development of reciprocity in exchanges and forms of collaboration. This exercise is useful not only for a more uniform flow of resources inside the network but also to strengthen relations, promote the sharing of governance rules and convergence on that meta-cultural dimension which gives meaning to being together and sharing a common enterprise at the service of the local community.

Despite the conceptual and methodological innovations that have been exper-imented in this empirical investigation, there are several limits that must be highlighted. First of all, it is well known that in the evaluation of network effectiveness, the study of the structure of relations in inter-organizational systems is essential but not enough [21]. There are many other aspects that should be taken into account in order to reconstruct a more complete picture of the governance network, such as the role of the stakeholders of the nodes and the point of view of the beneficiaries of the network actions. Moreover, one should be able to perform analyses of a more strictly qualitative nature, oriented to understand the meaning making of the various organizations of the network (cf. [15, 42]). In this way it would be possible to enter deeper into the network interactions between the nodes, as to bring out the "micro" processes that preside over structural transformations within the network. A second limit is in the cross-sectional character of the survey, which prevents or constrains the possibility of making statements that go beyond the descriptive dimension of the studied dimensions.

The proposed blockmodelling analysis, although it has a nature that is still purely "exploratory", is a technique that is able to make the characters of the network more "visible" [20]; however, in order to better appreciate the role of reciprocity, transitivity and homophily in the implementation of the network, as well as the effects of covariation between these parameters, it would be appropriate to test ERGMS models, now consolidated in the SNA's assets. Finally, from a cognitive point of view, it would be very useful to connect network effectiveness indicators with the performance of inter-organizational networks in terms of the impact of their action in the served community, although it is not easy to make adequate measurements of network outcomes in the frame of welfare community.

Acknowledgments Antonietta Riccardo thanks the Centro Universitario Cattolico that partially supported her activities. The authors thank Carmine Schiavone (director of the Caritas of the Diocese of Aversa), Nicola De Blasio (director of the Caritas of the Diocese of Benevento) and Angelo Moretti (director of the consortium "Sale della Terra") for their collaboration in the empirical phase of the research.

References

1. Provan, K.G., Milward, H.B.: A preliminary theory of interorganizational network effectiveness – a comparative study of 4 community mental health systems. Adm. Sci. Q. **40**(1), 1–33 (1995)
2. Anklam, P.: Net Work. A Practical Guide to Creating and Sustaining Networks at Work and in the World. Butterworth–Heinemann, Oxford (2007)
3. Krebs, V., Holley, J.: Building smart communities through network weaving. www.orgnet.com (2006)
4. Valente, T.V., Palinkas, L.A., Czaja, S., Chu, K., Brown, C.H.: Social network analysis for program implementation. PLoS One. **10**(6), e0131712 (2015). https://doi.org/10.1371/journal.pone.0131712
5. Kenis, P., Provan, K.G.: Towards an exogenous theory of public network performance. Public Adm. **87**(3), 440–456 (2009)
6. Raab, J., Kenis, P.: Heading toward a society of networks: empirical development and theoretical challenges. J. Manag. Inq. **18**(3), 198–210 (2009)
7. Provan, K.G., Kenis, P.: Modes of network governance: structure, management, and effectiveness. J. Public Adm. Res. Theory. **18**(2), 229–252 (2008)

8. Salvini, A.: Dentro le reti. Forme e processi della Network Governance. Riv. Trimest. Sci. Ammin. **4**, 39–58 (2011)

9. Bertin, G., Fazzi, L.: La Governance Delle Politiche Sociali in Italia. Carocci, Rome (2010)

10. Kazepov, J.: La Dimensione Territoriale Delle Politiche Sociali in Italia. Carocci, Rome (2009)

11. Wagner, A.: Reframing "social origins theory": the structural transformation of the public sphere. Nonprofit Volunt. Sect. Q. **IXXX**(4), 541–553 (2000)

12. Battistella, A.: La complessità delle reti sociali – 1. Prospettive Soc. Sanit. **38**(16), 6–11 (2008)

13. Battistella, A.: La complessità delle reti sociali – 2. Prospettive Soc. Sanit. **38**(17), 4–6 (2008)

14. Scardigno, F.P.: Programmazione e valutazione del servizio sociale nell'epoca del welfare di comunità. In: Salvini, A. (ed.) Dinamiche di comunità e servizio sociale, pp. 97–116. Pisa University Press, Pisa (2016)

15. Salvini, A., Gambini, E.: Fare rete. 15 linee guida per sperimentare la rete tra organizzazioni di volontariato. Cesvot, Firenze (2015)

16. Putnam, R.: Bowling alone: America's declining social capital. J. Democr. **6**(1), 65–78 (1995)

17. Bogason, P., Zolner, M. (eds.): Methods in Democratic Network Governance. Palgrave Macmillan, London (2007)

18. Sorensen, E., Torfing, J. (eds.): Theories of Democratic Network Governance. Palgrave Macmillan, London (2007)

19. Christopulos, D.: The governance of networks: heuristic or formal analysis? A replay to Rachel Parker. Pol. Stud. **56**(2), 475–481 (2008)

20. Robins, G., Bates, L., Pattison, P.: Network governance and environmental management: conflict and cooperation. Public Adm. **89**(4), 1293–1313 (2011)

21. Wang, W.: Exploring the determinants of network effectiveness: the case of neighborhood governance networks in Beijing. J. Public Adm. Res. Theory. **26**(2), 375–388 (2015)

22. Cross, R., Borgatti, S.P., Parker, A.: Making invisible work visible: using social network analysis to support strategic collaboration. Calif. Manag. Rev. **44**(2), 25–46 (2002)

23. Provan, K.G., Veazie, M.A., Staten, L.K., Teufel-Shone, N.I.: The use of network analysis to strengthen community partnership. Public Adm. Rev. **65**(5), 603–612 (2005)

24. Provan, K.G., Fish, A., Sydow, J.: Interorganizational networks at the network level: a review of the empirical literature on whole networks. J. Manag. **33**(3), 479–516 (2007)

25. Neal, Z.P.: A network perspective on the process of empowered organizations. Am. J. Community Psychol. **53**, 407–418 (2014)

26. Wang, R., Tanjasiri, S.P., Palmer, P., Valente, T.W.: Network structure, multiplexity, and evolution as influences on community-based participatory interventions. J. Community Psychol. **44**(6), 781–798 (2016)

27. Metallo, G., Cuomo, M.T., Tortora, D., Galvin, M.: Innovation networks and social enterprises. A social network analysis of the third sector in Italy. Riv. Piccola Impresa. (3), 1–24 (2016)

28. Delle Cave, L.: Forme, dinamiche e reti sussidiarie del Terzo Settore nella realtà napoletana. Riv. Impresa Soc. **1**, 38–51 (2013)

29. Borgatti, S.P., Everett, M.: Models of core/periphery structures. Soc. Networks. **21**, 375–395 (1999)

30. Burt, R.: Structural Holes. The Social Structure of Competition. Harvard University Press, Cambridge (1992)

31. Corbisiero, F.: Terzo settore e sussidiarietà territoriale. In: Musella, M. (ed.) La Sussidiarietà Orizzontale, pp. 203–217. Carocci, Rome (2013)

32. Salvini, A.: Introduzione. Trionfo, declino e nuove prospettive di sviluppo del volontariato in Italia. Sociol. Ric. Soc. **XIII**(96), 9–31 (2011)

33. Vasca, F., Riccardo, A., Capuano, G. (eds.): Reti di periferia. Sistemi sociali virtuosi fra Terra di Lavoro e Terra dei Fuochi. Aracne Editore, Ariccia (2016)

34. Vasca, F., Tangredi, D., Riccardo, A., Iervolino, R.: Modelli di costruzione di comunità tra coopetizione e cooperosità. In: Salvini, A. (ed.) Dinamiche di comunità e servizio sociale, pp. 117–139. Pisa University Press, Pisa (2016)

35. Negrón, R., Leyva, B., Allen, J., Ospino, H., Tom, L., Rustan, S.: Leadership networks in Catholic parishes: implications for implementation research in health. Soc. Sci. Med. **122**, 53–62 (2014)
36. Leonard, R., Bellamy, J.: Dimensions of bonding social capital in Christian congregations across Australia. Voluntas. **26**, 1046–1065 (2015)
37. Everton, S.F.: Networks and Religion. Ties That Bind, Loose, Build-Up, and Tear Down. Cambridge University Press, Cambridge (2018)
38. Granovetter, M.: Problems of explanation in economic sociology. In: Nohria, N., Eccles, R.G. (eds.) Networks and Organizations: Structure, Form and Action, pp. 25–56. Harvard Business School Press, Boston (1992)
39. Jones, C., Hesterley, W.S., Borgatti, S.: A general theory of network governance: exchange conditions and social mechanisms. Acad. Manag. Rev. **22**(4), 911–945 (1997)
40. Crossley, N., Bellotti, E., Edwards, G., Everett, M.G., Koskinen, J., Tranmer, M.: Social Network Analysis for Ego-Nets. Sage, London (2015)
41. Doreian, P., Batagelj, V., Ferligoj, A.: Generalized Blockmodeling. Cambridge University Press, Cambridge (2005)
42. Salvini, A., Psaroudakis, I.: Towards innovative practices of networking among volunteer organizations. In: Borghini, A., Campo, E. (eds.) Exploring the Crisis. Theoretical Perspectives and Empirical Investigations, pp. 89–104. Pisa University Press, Pisa (2015)

Index

© Springer Nature Switzerland AG 2020
G. Ragozini, M. P. Vitale (eds.), *Challenges in Social Network Research*,
Lecture Notes in Social Networks, https://doi.org/10.1007/978-3-030-31463-7